D0846867

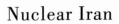

Nuclear Iran

Nuclear Iran

Jeremy Bernstein

Harvard University Press

Cambridge, Massachusetts, and London, England

2014

Copyright © 2014 by the President and Fellows of Harvard College

ALL RIGHTS RESERVED

PRINTED IN THE UNITED STATES OF AMERICA

First printing

Library of Congress Cataloging-in-Publication Data

Bernstein, Jeremy, 1929– author.
Nuclear Iran / Jeremy Bernstein.
pages cm
Includes bibliographical references and index.
ISBN 978-0-674-41708-3 (alk. paper)
1. Centrifuges—Iran—History.   2. Nuclear reactors—Iran.
3. Uranium—Isotopes.   4. Plutonium—Isotopes.   5. Nuclear
weapons—Iran.   I. Title.
TP159.C4B47 2014
623.4'5119—dc23   2014010217

Contents

PART III   Dual Use

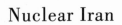

Nuclear Iran

Prologue

IN THE SUMMER OF 2009 I happened to watch a television interview with an Israeli general. He was identified as the head of Israeli intelligence. He struck me as an impressive and serious man. He was of course asked about Israel's concern with its Arab neighbors—Syria and the like. Then the subject turned to Iran. As the general made clear, the presence of nuclear weapons in Iran is for Israel an existential matter—to have a nuclear armed near neighbor whose state policy appears to be the destruction of Israel is intolerable. The general was then asked what the Israelis would regard as the trigger point—the line in the sand. He replied that it would be when the Iranians acquired the "knowledge" to make nuclear weapons. This statement appalled me.

After all, how would one learn about another country's state of "knowledge"? The Iranians, who have consistently claimed that they are not building

FIGURE P.1. A centrifuge plant.
(United States Department of Energy)

weapons, are not going to inform the Israelis. In any event, the Iranians surely do have the knowledge, if for no other reason than that they bought the plans for a nuclear weapon from the Pakistani proliferator A. Q. Khan. These are presumably the same plans for a Chinese-designed weapon that Khan sold to Libya. These were turned over to us by the Libyans. No, knowledge is not the issue. The issue is the materials needed to make a weapon—the highly enriched uranium and plutonium. If the Iranians accumulate enough of these to make a weapon, then I think, despite what they say, they will make one. This is the line in the sand.

After listening to the general it struck me that there was something I could do. I could write a book that would make this clear. I have already written two books on nuclear weapons. The first was on the element plutonium.[1] It focused on the history of its discovery and on its very bizarre physical and chemical properties as well as the role it played in the Manhattan Project. Of Iran, there is no word. The second book was an attempt to explain what nuclear weapons are and how they work.[2] The Iranian program gets a brief mention. For example, I noted that centrifuges play a role but I never explained exactly what they were and how they worked. In my

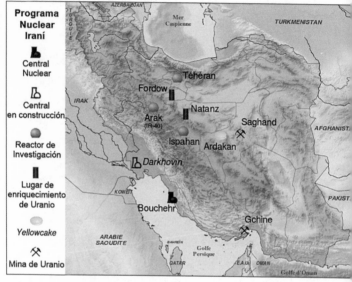

FIGURE P.2. Map of the Iranian nuclear sites. "Central" is the Spanish for "power plant." Unlike many other maps, this one shows the location of the uranium nuclear mines as well as the sites where uranium is enriched.

(Wikimedia Commons)

plutonium book I did not discuss the reactor program in Iran and what it means for their nuclear weapons. In this book, matters like this will be the main focus.

I have divided this book into three parts—"Uranium," "Plutonium" and "Dual Use." In the first part I discuss the history of the sort of centrifuges used to produce the weapons-grade uranium needed for bombs. One of the most bizarre aspects of this is that these centrifuges were designed by German prisoners of war in the Soviet Union. One may well ask, how did this information get from Sochumi on the Black Sea to Tehran? I explain. I also describe in some detail what is known about the Iranian centrifuges and the Iranian enrichment program in general. In the second part I discuss the completely different technology required to produce plutonium and to use it in a nuclear weapon. Here we deal with reactors and implosion. I have tried to intersperse the technology and the human story. In the last part I give my assessment of what the future holds. As I write this, that assessment seems to be changing daily as new facts about the Iranian program are revealed and as negotiations proceed. But it is essential that we understand these facts.

Many years ago I spent a lot of time interviewing Stanley Kubrick for what became a *New Yorker* profile. He told me how he became interested in nuclear weapons after the Cuban missile crisis. It struck him that most people were not at all interested as soon as the crisis passed. He said that they had less interest in nuclear weapons than they had in city government. That may have been true in the early 1960s when I did my interviews, but it is not true now. Thanks to the Iranians and North Koreans, nuclear weapons are something we can no longer ignore. I think those of us who have some understanding of the subject have a responsibility to share it.

PART I

Uranium

I

Round and Round

IN 1913 IN THE COURSE of a conversation with her cousin, the British radiochemist Frederick Soddy, the Scottish doctor Margaret Todd coined the term "isotope." Soddy had found persuasive evidence that chemical elements could come in varieties that had the same chemical properties but had different masses. Therefore all these varieties would occupy the same position in the periodic table. Soddy did not know what name to give to this phenomenon. His cousin suggested from the Greek -isos—the same—and topos—place or position. In 1921 Soddy was awarded the Nobel Prize in Chemistry for his work. Miss Todd had died in 1918 shortly after her biography of her lifelong partner Sophia Jex-Blake was published.

Soddy's methods for discovering isotopes were largely chemical. But just after the First World War the British physicist and chemist Francis William

Aston produced a new method that was basically physical and revealed isotopes directly. It was well known that charged particles in evacuated tubes could have their orbits diverted by electric and magnetic fields. Indeed, by combining the two fields one could make these "ions" follow parabolic orbits. If one had only a magnetic field, these parabolas degenerated into circular arcs. The lighter ions get deflected more than the heavier ones if the two ions have the same electric charge. This is the basic idea of the so-called mass spectrometer, which was Aston's invention although the British physicist J. J. Thomson had come up earlier with a simpler version of it. The Canadian American physicist Arthur J. Dempster independently invented a similar device. In 1935 he separated two of the isotopes of uranium—uranium 235 and uranium 238. Using his device Aston separated the isotopes of dozens and dozens of elements. For a given element the different isotopes show up as distinct lines on a photographic plate. From the intensity of these lines one can even tell which isotopes are more common than others. To take a case in point, Aston found that for neon two lines showed up, which revealed two of the three stable isotopes.

Aston, who was awarded the Nobel Prize in Chemistry, died on November 20, 1945. This meant that

FIGURE 1.1. The centrifuge used in Stanley
Kubrick's film *2001—A Space Odyssey*.
(Licensed by Warner Bros. Entertainment, Inc. All Rights Reserved.)

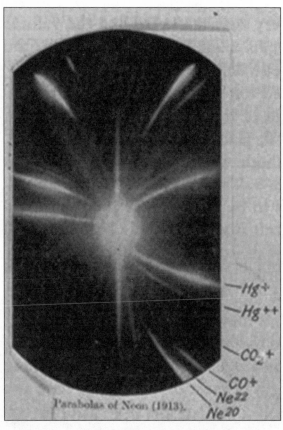

—Hg^+

—Hg^{++}

—CO_2^+

—CO^+

Ne^{22}

Ne^{20}

Parabolas of Neon (1913).

FIGURE 1.2. Mass spectrograph photo showing at the bottom right the two lines from the isotopes of neon.

(Wikimedia Commons)

he lived long enough to learn about atomic bombs. One wonders whether he knew that descendants of his mass spectrograph had been used at Oak Ridge, Tennessee, to help separate the isotopes of uranium so as to produce the explosive material for the bomb that flattened Hiroshima. Dempster lived until 1950 and indeed during the war worked on the Manhattan Project.

During the First World War Aston had gotten to know another physicist, Frederick Lindemann. Both men were working on the physics of airplane design. Lindemann developed a theory of aircraft spin recovery and learned to fly so he could test it, which he did successfully. At the time Lindemann did not believe in isotopes. Ironically, he and Aston published a joint paper in 1919 in which, for the first time, the idea of using a centrifuge to separate the isotopes of gasses was suggested.[1] This paper was the inspiration for all the later work.

Before I turn to the contents of their paper, I want to say a bit more about Lindemann. He was born in 1886 in Germany of an American mother and a father who had already emigrated to England. Lindemann was a very good tennis player and was actually playing in a tournament in Germany when the First World War broke out. He later competed at Wimbledon.

He was also a teetotaler and a vegetarian. He took an appointment at Oxford at about the same time that Aston took one at Cambridge. In the Second World War he became Churchill's principal science advisor. In 1941 he became Baron Cherwell, which inspired one of the great British academic doggerel verses:

> Lord Cherwell, when the war began,
> Was plain professor Lindemann.
> But now, midst ministerial cheers,
> He takes his place among the peers.
> The House of Christ with one accord
> Now greets its newly risen Lord.

It helps to know that Christ Church was the name of the Oxford college to which Lindemann belonged and Cherwell is the name of a nearby stream.

Aston and Lindemann begin their paper by discussing a problem that on its face has nothing to do with centrifuges. They ask if the gravitational attraction of the Earth can separate the isotopes of the gasses in its atmosphere. The answer is yes, and here is how it works. Let us for the sake of argument suppose that the atmosphere consists of a single gas. I will take this to be neon, since what Aston and Lindemann studied was whether gravity would separate the isotopes of neon in the stratosphere. Let me begin by supposing that neon has no isotopes. We will add the isotopes later.

Let us imagine a very small boxlike volume way up in the stratosphere and ask what forces act on the neon gas located within it. The first thing that comes to mind is gravity, which attracts all the neon molecules down toward the surface of Earth. Why don't they all simply fall down and collect on Earth's surface? This is because there is a second force that pushes them up. This is the collective force exerted by all the molecules pressing on the bottom of the box and exerting a pressure. (One could ask the same question about clouds. What keeps them up?) When the two forces are equal, we have a state of equilibrium. Now suppose neon has an isotope that is heavier than the one we have been considering. We can make the same argument, but since the individual molecules are heavier, it requires fewer of them to achieve equilibrium. The pressure pushing them up is the same for the two isotopes, but the force of gravity is greater for the heavier one, hence it takes fewer of them to balance the pressure. Therefore, the heavier isotopic gas is less dense. If the two isotopes are in the box together, then, all things being equal, there will be a very small difference in the densities of the two components, and that is what we will try to measure.

If we compute the ratio of the two densities under reasonable assumptions, it has an exponential behavior. What counts is the exponent that determines how the exponential grows. Let us call the two densities m_1 and m_2—the densities are simply these masses per unit volume. Then the exponential that determines the mass ratio is proportional to the mass difference Δm multiplied by the height h above the earth times the gravitational acceleration g that any molecule experiences as a consequence of the Earth's gravity. Thus, the argument of the exponential that determines the ratio is proportional to the product Δmgh. Whether the density ratio is greater or less than 1 depends on the sign of Δm. Let me point out that once equilibrium has been established, the density ratio is fixed and will remain unchanged so long as the equilibrium is maintained. We do not increase the degree of separation as time passes. We shall keep this in mind when we turn to centrifuges. Aston and Lindemann proposed to take a balloon that could collect samples of the ambient atmosphere and let it rise to about 100,000 feet. They thought that there would be a small but measurable effect on the isotopic densities. As far as I know, this experiment was never performed and hasn't been even to this day.

Those of you who remember Stanley Kubrick's film *2001* will recall one of the astronauts jogging around a centrifuge that was located on a spaceship in outer space. As I can testify, having visited the set several times, this was a real centrifuge built at the cost of about $300,000 by the Vickers Engineering Group. It had a 38-foot diameter and a maximum peripheral speed—the speed a stationary object located on the periphery of the centrifuge would have as observed by an observer located at rest in the center of the centrifuge—of about three miles an hour. Kubrick told me that he was thinking of selling rides to help recover the cost. He asked me not to put that in the profile I was writing of him. It is amusing to note that to reproduce the gravitational acceleration of 32 feet per second per second at the periphery of this centrifuge, a peripheral speed of about seventeen miles an hour would be required. As it was, it produced about one foot per second per second.

Aston and Lindemann did not live long enough to see Kubrick's film, but in 1919 they grasped the analogy between the centrifugal force in the centrifuge and the action of gravitation when it came to separating isotopes. The acceleration g is replaced by the acceleration that is directed toward the perimeter of the centrifuge. If v is the speed of an object located at a distance r from the center, then this acceleration is v^2/r. Kubrick's centrifuge produced an acceleration of his astronauts of about a half a foot per second per second, whereas the gravitational acceleration at the surface of the Earth in the same units is about 32 feet per second per second. A good modern centrifuge can produce accelerations at the periphery a thousand times greater than this. It is customary to rewrite the peripheral velocity in terms of frequencies of rotation. Suppose an object located at a distance r from the center makes a complete rotation f times a second. Because the distance it goes around each time is just the circumference of a circle of radius r—that is, $2\pi r$—then v is given by $2\pi fr$. It is customary to define the angular frequency ω as $\omega = 2\pi f$. Thus, in terms of ω, the centrifugal acceleration is given by $\omega^2 r$. Now let us consider a gas centrifuge.

In the relevant examples, these consist of long thin cylinders into which the gas is inserted. These cylinders can be made of various metals or, in the recent versions, carbon fiber. We shall discuss the details in another chapter. The cylinders are mounted vertically on some sort of bearing. They are rotated at high speeds around the central axis. After they are set rotating, the gas is inserted. The first thing that comes to mind is the thought that the cylinders would spin while leaving the gas in place. This would be true if the materials of the cylinders were perfectly smooth. But they have tiny rugosities off which the gas molecules can bounce and in which they can stick before bouncing. The molecules then partake of the rotation, and when they return to the body of the gas, they convey the sense of rotation to other molecules with which they collide. Soon the entire body of gas is rotating. Once this happens the centrifugal force impels the gas molecules toward the periphery of the cylinder. But this force is opposed by a counterforce due to the pressure of some of the gas molecules. When these forces are equal, we have an equilibrium situation that is analogous to the one we have already discussed where gravity played the role of the centrifugal force. It is important to note that, like the atmospheric case, once equilibrium has been established between the centrifugal and pressure forces the degree of separation of the isotopes remains the same. It does not matter how long we allow the gas to spin—the degree of separation does not change. This has profound consequences for the practical separation of isotopes.

There is a difference between the centrifugal force case and gravitation. In the centrifugal force case the acceleration depends on where we are in the cylinder. The farther away from the center, the stronger the acceleration. If we want to know what is the best separation the centrifuge can produce, we will take this distance as big as we can, which is the radius of the cylinder. The speed of a spot at this radius is called the "peripheral speed," v_p. Once again we can do the same mathematics that led to the answer in the gravitational case. We find as before that the ratio of the densities of two isotopes separated by the centrifugal force is given by an exponential whose exponent in this case is proportional to $\Delta m \, \omega^2 r^2 = \Delta m v_p^2$. In a good centrifuge, v_p is of the order of 355 meters per second or greater. This is somewhat larger than the speed of sound in air. (Kubrick's centrifuge did about 1.5 meters per second.) If we put in all the numbers, then in a typical case the exponential is something like .06, which is pretty small, indicating that a single centrifuge has limited separation capacity.[2] This is the kind of number we have to deal with if we try to use the centrifuge to separate the isotopes of uranium—a case that will be our principal concern. In considering the use of centrifuges, Lindemann and Aston concluded, "Separation by this method therefore seems possible though difficult and costly."[3] So it turned out to be. In fact it was not done until 1934, when Jesse Beams of the University of Virginia managed to separate two isotopes of chlorine gas—chlorine 35 and chlorine 37—using a centrifuge. None of this work was done with any military applications in mind. And then came the discovery of fission.

In December 1938 the German radiochemists Otto Hahn and Fritz Strassmann discovered nuclear fission. This statement must be qualified. They in fact observed something of which they had no understanding that turned out to be the result of nuclear fission. To understand this we must back up a little. In 1932 the British physicist James Chadwick discovered the neutron. It was accepted at the time that the atomic nucleus consisted of charged particles (protons) and electrically neutral particles (neutrons). The assumption was that the neutron was an electron and a proton bound together. Indeed, that is what Chadwick thought and he said so in his paper. But it soon became apparent that the neutron was a particle in its own right. It also became apparent that the neutron, because of its lack of electric charge, was the perfect candidate to penetrate atomic nuclei. Several groups began bombarding a variety of materials with neutrons. One of the most successful was a group led by Enrico Fermi in Rome. These people bombarded every element they could lay their hands on, finally coming to uranium, which was the heaviest element then known. Fermi's expectation was that the neutrons would transform uranium into even heavier elements— "transuranics." In fact that is what he thought had happened, and he even published the result. It turned out that he was wrong, and if some extra shielding had not been added to his equipment he would have discovered fission.

There was a German group that was working in Berlin. One of its members was the aforementioned Otto Hahn. He had had a longtime, on-and-off collaboration with an Austrian-born physicist, Lise Meitner. After the neutron was discovered they joined forces again. They were soon joined by Strassmann. He was an outspoken and very courageous anti-Nazi, which meant that he was virtually unemployable in Germany. But Hahn and Meitner found some money to pay him as an assistant. The three of them went in search of the transuranics. Meitner had been born Jewish but had converted so as to blend in better with the German cultural scene. She thought that her Austrian citizenship would protect her, in any event. It did until 1938 when the Germans annexed Austria and Meitner found herself to be a German citizen subject to all the Nazi race laws. Her life was in danger, and she was lucky to be able to escape to Holland and even luckier to leave Holland for Sweden before the German occupation. There she spent the war. She had some contact with Hahn, and in December 1938 he informed her by letter of the incomprehensible results he and Strassmann were getting. As I mentioned, they were expecting to find traces of an element beyond uranium in the periodic table. Instead they were finding barium, which is somewhere in the middle of the table. Hahn could find no explanation for this and asked Meitner's help.

Meitner was going to spend the Christmas vacation in a friend's house in Kungälv on the west coast of Sweden. Her nephew Otto Frisch, a Viennese-born physicist, was to join her. Frisch had been in Berlin but had left in 1933, first to England and then to Niels Bohr's Institute for Theoretical Physics in Copenhagen. He had some physics problems he wanted to discuss with his aunt, but she would have none of it. All she wanted to do was to discuss Hahn's letter. Frisch suggested that it might have been a simple mistake. Meitner would have no part of that either. Hahn didn't make mistakes like this. If he said there was barium, then there was barium. They went for an outing in the woods, Frisch on cross-country skis and Meitner trotting alongside. What happened—at least as Frisch recounted it later—is one of the great folklore tales of twentieth-century physics.[4] How much of the story is literally true, I do not know.

Somewhere during this outing they had an epiphany and understood everything. This part I believe. But then Frisch describes their sitting on a tree trunk and calculating things on scraps of paper using some pencils they had brought along. Did they have slide rules for the arithmetic? Did they do the arithmetic in their heads? This is the part that I find difficult to take literally. The only account is Frisch's. But what was the point? To understand this we have to present some ideas about nuclear structure that were then in use—and still are.

The nucleus of a heavy element contains dozens of particles—neutrons and protons, or "nucleons," to use the generic term. For example, the isotope uranium

238 has 238 nucleons in its nucleus, of which 92 are protons. As a practical matter it is not possible to treat these nucleons individually. Instead what one does is study their collective behavior. One model that is used is what is known as the "liquid drop." A water drop, for example, consists of a myriad of water molecules held in a volume by the electrostatic forces that act on the molecules on the surface. A nucleus has a roughly spherical shape, and one can think of the nucleons on the surface as being held there by the nuclear forces. This suggests a picture of what might happen if a neutron impinges on such a nucleus. It might set up vibrations in the "drop" that could cause it to change its shape. It would get distended. If the energetics were right (I come back to this shortly) the drop might break up into droplets—in this case, nuclei of smaller mass and possibly neutrons as well. Frisch and Meitner were sure that this is what must have happened in the Hahn-Strassmann experiment. One of the "droplets" was barium, and by adding up the electric charges they concluded that the other must have been krypton. Krypton is a chemically inert gas, which meant that in their experiment it simply floated off. They would also not have been able to

observe any of the emitted neutrons. This was the picture, and now Meitner and Frisch had to show that it was energetically possible.

Einstein had taught us that there is a relation between mass and energy—$E = mc^2$—where c is the speed of light propagating in a vacuum. To make the process they were postulating energetically possible, the mass of the uranium nucleus had to be greater than the sum of the masses of the barium and krypton nuclei. To show that this worked, Frisch and Meitner had to know these masses. As it happened, Meitner had had a research associate in Berlin named Carl Friedrich von Weizsäcker. His father, Ernst von Weizsäcker, was a very high official in the foreign office. When Meitner was in danger, she had appealed to him, but he never responded. He was later convicted of crimes against humanity at Nuremburg. Young Weizsäcker was a brilliant theoretical physicist. He had devised a formula using the liquid drop model for computing—at least approximately—the masses of all but the lightest nuclei. The formula contained parameters that were chosen to give the best fit. Frisch reports that his aunt knew all these numbers by heart. There are various versions of this formula, and it isn't known which one they used. They found that the masses worked out. In fact the mass difference was so substantial that the "drops" would acquire energies orders of magnitude greater than those produced in chemical reactions. With this knowledge Meitner returned to Stockholm and Frisch to Copenhagen to begin writing up their discovery.

In Copenhagen Frisch had three tasks. He needed to find a name for the process, he needed to do an experiment to show that the "droplets" actually had the predicted high energy, and he needed to tell Bohr about it. He asked a biologist colleague what they called cell division in biology and was told that they called it "fission," so fission it became. The first Frisch-Meitner paper, which came out in February a few weeks later, is the only place I know where the term "fission" is in quotation marks. Frisch performed his experiment and saw the electric pulses that the fission fragments made. He of course told Bohr.

I had a chance to talk to several physicists, people like Hans Bethe, who were around when the discovery of fission was made. All of them reacted in the same way. How could they have missed it? It was totally obvious to them once it was explained. Bohr had the same reaction. As it happened he was about to leave for the United States, where he was going to spend time in Princeton. Accompanying him was his assistant and amanuensis, the Belgian-born theoretical physicist Léon Rosenfeld. On the way over Bohr had a blackboard set up in his stateroom and he and Rosenfeld went over the fission calculations. Bohr had an understanding with Frisch and Meitner not to say anything in public until their paper was published. But he had forgotten to tell Rosenfeld about it. Bohr stayed in New York while Rosenfeld went directly to Princeton, accompanied by John Wheeler, who had met them at the boat. Rosenfeld told Wheeler on the train ride to Princeton. The night that Rosenfeld arrived happened to be the night when the Princeton physicists had their journal club meeting. They shared the latest news in physics. Rosenfeld came and spilled the beans. Very rapidly his news spread all over the United States. Experiments were begun immediately in several laboratories. Soon the missing fission neutrons were observed and it became clear that chain reactions were possible, because these would be induced by the newly created neutrons. It also became clear what the implications of this could be.

Bohr and Wheeler began work on what would be their monumental article on fission. Bohr gave a lecture, and in the audience was an acerbic Czech-born physicist named George Placzek. When Bohr was finished, Placzek told him that the whole thing was total nonsense. Quantum mechanics predicted that the less energetic the impinging neutron, the higher the fission reaction rate would be. This has to do with the wave nature of the neutron. The slower the neutron moves, the bigger the wavelengths. This improves the overlap between the neutron and the uranium nucleus, and hence the fission rate is enhanced. But Placzek complained that none of this made sense from the point of view of the energy. Where did the energy come from for the neutron to initiate the vibrations that cause the nucleus to fission? This was a very good question, and Wheeler told me that Bohr was very troubled by it. He and Wheeler started walking randomly around Princeton, and Bohr suddenly came up with the answer— and it changed the application of fission to create nuclear energy forever after. We need to say some more about the nuclear force.

The main point to be made is that the nuclear force has a very short range. The nucleons have to be practically sitting on top of each other for it to take effect. Keep this in mind. Let us begin with uranium 238. When it absorbs a neutron, it becomes temporarily uranium 239. Compare this with the relatively rare isotope uranium 235. When it absorbs a neutron it becomes uranium 236. In this case all the nucleons have partners to share forces with—unlike uranium 239, where there is an odd nucleon out. This means that uranium 236 is more tightly bound than uranium 235, which also has an odd nucleon out, while the opposite is true of uranium 238 and uranium 239. The more tightly bound the nucleus is, the less its mass. The reduction in mass is converted into energies, which are different in the two cases. This difference makes all the difference. In the one case there is an energy release that exceeds the barrier energy, which has to be overcome to produce fission when the compound nucleus is formed; and in the other case the energy release does not exceed the barrier energy. Nuclei with the former property are called "fissile." They can be fissioned by neutrons of any energy. Bohr realized that it was the very rare isotope uranium 235 that was responsible for any fission that had been observed. This gave him a sense of relief. To make a nuclear weapon, for example, would require the monumentally difficult task of separating isotopes on an industrial scale, which he thought would never happen since it would require the resources of an entire country. He was wrong on both counts.

It did happen, and although it required resources that only the United States could then muster, it hardly took the entire country. The facility at Oak Ridge, Tennessee, where this was done cost hundreds of millions of dollars and employed during the war some 70,000 people, but we had plenty of resources to spare.

2

Frisch, Peierls, and Dirac

IN THE SUMMER OF 1939 Frisch left Copenhagen for what he thought would be a short trip to Birmingham. The war broke out and he stayed in Birmingham. There he found another refugee from Germany, Rudolf—later Sir Rudolf—Peierls, who had been a professor at the University of Birmingham since 1937. He was in England when Hitler came to power and he received permission to stay. At a conference in 1930 in the Soviet Union he had met the Russian physicist Yvgenia Kanegisser. He married her. "Genia" Peierls was, as I can testify, an irresistible force. In her presence the "prof" often maintained a discreet silence. Early in their marriage—and I see Genia's hand in this—they decided that extra living quarters in any house they had should not go to waste. So itinerant physicists took up lodging. It is quite a list. In the early 1930s it included people like Hans Bethe. After the war

Freeman Dyson was a temporary lodger. For part of the war, until he went to Los Alamos in 1943, Klaus Fuchs was a resident. Genia complained that he practically never talked. He was, she said, like one of those machines you had to put a coin in for it to make music. Until 1950, when he was arrested, neither of the Peierls knew that Fuchs was a Soviet spy. In 1939 the resident was Frisch.

Frisch and Peierls were probably more aware than most of the menace of Nazi Germany. They also knew that even though many of the best scientists had been forced out, there was still a formidable group that had remained—people like Werner Heisenberg. They also knew that through popular articles there was a wide awareness of how powerful a force fission could in principle be. They were persuaded that if atomic weapons were going to be built, the Germans could not be allowed to have a monopoly. Both of them had the status of enemy aliens in Britain and so had no knowledge of anything the British establishment might be doing. They decided to study the matter themselves.

Bohr's paper on the role of uranium 235 had been published in the open literature, as had the fact that in the fission of uranium 235 neutrons are also emitted, which had first been suggested by Frisch and

Meitner. One should understand that the fission
that Hahn and Strassmann observed was not unique.
It was not even the most probable. Some sixty different isotopes can be produced by uranium fission.
They saw barium because their setup revealed barium. In any event, on average over all the different
possibilities about 2.5 neutrons are emitted per fission. That is to say that, for example, in some pro-
cesses three are emitted, while in others two are
emitted. Frisch and Peierls knew this number at
least approximately. They then performed a thought
experiment.

They imagined that there was a process that would
produce as much uranium 235 as desired. How much
would be needed to make a nuclear explosion? What
would be the minimum amount—the "critical mass"?
If this turned out to be tons, then the whole idea
could probably be dismissed as an impractical chi-
mera. They decided to take this hypothetical chunk
of uranium in the shape of a solid sphere. Not only
would this simplify the geometry, but for reasons I
will explain it could minimize the mass. Of all the
solids, a sphere has the least surface area for a fixed
volume. For example, if you take a cube of the same
volume, its six faces have more surface area. We shall
soon see the reason why this matters.

Let us imagine that a neutron is introduced into the sphere and that it fissions a uranium nucleus, producing other neutrons along with the fission fragments. Let us follow the path of one of these neutrons. It will collide with another uranium nucleus. The most likely thing to happen is that it will simply bounce off this nucleus. On average it takes about five of these "elastic" collisions before another fission takes place. The first question we want to answer is this: How far, on average, does such a neutron travel before a second fission is produced? This is called the "mean free path" for fission. This depends on two more questions: How many uranium 235 nuclei are to be found in a cubic centimeter of the stuff? And how likely is it that a collision will produce fission? The first of these Frisch and Peierls had no trouble answering.

The mass density of uranium, ρ, is about 19 grams per cubic centimeter. It is a very dense substance — about 1.6 times denser than lead and about 19 times denser than water. Each uranium nucleus weighs about 3.9×10^{-22} grams, so the number in a cubic centimeter is about $19/3.9 \times 10^{23}$, or about 5×10^{23} uranium 235 nuclei per cubic centimeter—a huge number. Let us call it n. The second question, about the likeliness of fission, they had no way of answering, at least empirically. Whatever data there was would not be shared with them. They had to guess. To explain their guess I need to tell you how physicists characterize such a likelihood. A homey example may help. Suppose there is the side of a house with one glass window and I want to throw baseballs in such a way that I break the window. It is clear that, all things being equal, my chances improve, the bigger the area the window has. Physicists measure the likelihood of a reaction in terms of an effective area—a "cross section." Because of quantum mechanics this area is not generally related to the physical size of the target, but it is a good way of describing these likelihoods. If we call the cross section for fission σ_f, then the mean free path for fission l_f is equal to $1/n\sigma_f$. In the absence of empirical data Frisch and Peierls put in the largest value for σ_f that quantum mechanics allows. This gave them a value of 2.6 centimeters for the mean free path. The actual value is 16.5 centimeters. This led them, as I shall explain, to a gross underestimation of the critical mass.

This is not the place to describe the mathematics of this exercise. In fact, in the two short notes they wrote in 1940 there is almost no mathematics.[1] They only give the results. We can get a feeling for them by the following line of reasoning. If we make the radius of the uranium sphere smaller than the mean free path, then the chances are that a neutron will escape from the surface before it can cause an additional fission. That is why we want to minimize the surface area. If we make the radius larger than the mean free path, then we take the risk of having more mass than we need. So let us take the radius at just the mean free path. The mass is then the volume of the sphere multiplied by the density of the uranium: $m = 4/3\pi\, l_f^3\, \rho$. If we use their number for the mean free path, we find that the mass is about 1.4 kilograms. Doing the calculation more carefully, they found the critical mass to be only 600 grams, a little over a pound. This seemed to them very promising. They felt that it should not be impossible to separate a little over a pound of uranium 235. We might wonder what they would have thought if they had known the correct numbers and found something closer to the correct answer of about 46 kilograms—a little over a hundred pounds. (If they had used a cube, because of the larger surface area they would have found a result 1.24 times larger. The number one often finds, about 52 kilograms, refers to uranium that is highly enriched—over a 90 percent concentration of uranium 235.) Would they have given up trying to build a bomb? They recognized that to make an explosion you need more than the critical mass. Just at the critical mass the chain reactions start but then die out. You need "supercritical" masses. For example, the Hiroshima bomb, which used a mixture of about 90 percent uranium 235 and 10 percent uranium 238, used about 60 kilograms of uranium 235. What Frisch and Peierls proposed was to make two hemispheres of subcritical uranium and then to rapidly bring them together to make a supercritical mass.

One may still wonder if this would produce an explosion. An explosion is produced only when large amounts of energy are produced in a short time. Is that true here? We can give a qualitative argument along the following lines. The neutrons produced in fission are very energetic. They move with speeds about a tenth of the speed of light. I will take this to be 10^9 centimeters a second. To make things simple I will take the mean free path to be 10 centimeters. Hence, it takes about 10^{-8} seconds between fission events in uranium 235. This time interval was given the name "shake" at Los Alamos. Let us consider a kilogram of uranium 235. We may ask, How many uranium 235 nuclei are there in a kilogram? We can use the same sort of argument we used to find the number in a cubic centimeter to learn that there are about 2.6×10^{24} in a kilogram. Now let us suppose that each fission produces two neutrons. So first there will be two, and then four, and then eight, and so on. This is what a chain reaction does. After n such generations, 2^n neutrons will be produced. We can now ask, After how many generations do we produce enough neutrons to fission the kilogram? So we equate 2^n to 2.6×10^{24} and solve for n. We get about $n=80$. But each generation takes a shake. So it takes about 80 shakes—about a microsecond—to fission a kilogram. We do indeed have an explosion! Frisch and Peierls put all this and more in the two reports they wrote. But then what to do with them?

The chairman of the physics department at the time was an Australian-born physicist named Marcus Oliphant. He had been responsible for bringing Frisch and Peierls to Birmingham in the first place. He understood their citizenship situation. Peierls once told me that at the time Oliphant occasionally would come to him and say, "Have you ever seen an electromagnetism problem like this one? Do you have any idea of how to solve it?" Peierls knew at once that the problem was related to radar, but he and Oliphant maintained the fiction that it was just another physics problem. They brought their two papers to Oliphant. He was very impressed. He got the papers to Henry Tizard, a physical chemist who was an advisor to Churchill on radar. A committee was formed called the "MAUD Committee" to study the matter. The name has an amusing origin. At the time the committee was being formed, a communication from Bohr arrived. It ended with the mysterious phrase, "AND TELL MAUD RAY KENT." They were sure that Bohr must be trying to convey something of great significance and they tried desperately to break the code. It was only in 1943, when Bohr escaped to England, that they learned that Maud Ray of Kent had been governess to Bohr's children during one of their visits to England. In April 1940 the six-person committee had its first meeting. Frisch and Peierls were made consultants.

By this time Peierls had gotten some cross-sectional data from the United States and had raised his critical mass estimate to 12 kilograms. Frisch was primarily an experimental physicist, and he began trying to make an actual separation of the isotopes. He tried to make use of a method that had been invented in Germany by Klaus Klusius. The idea was to put a hot wire in the middle of a glass tube containing a gas with the different isotopes. The theory predicted that the lighter isotope would collect on the wire while the heavier isotope would collect on the cooler glass. Frisch had the idea that about 100,000 such tubes would do the trick, but he needed gaseous uranium. The form that is universally used is uranium hexafluoride, known colloquially as "hex." Its molecules are six fluorine atoms attached to one uranium atom.

It is a very nasty substance. At room temperature it is a solid grey crystal that can interact violently with water. At higher temperature it becomes a gas that is very corrosive with metals. Frisch had no hex, but James Chadwick, who was then in Liverpool, did. Frisch found Chadwick a little strange. His head moved from side to side like a bird's as Frisch scrutinized him. Finally Chadwick simply asked how much hex Frisch wanted and even invited him to do his experiments at Liverpool. Frisch discovered that using the gaseous form of hex, the Klusius method did not work sufficiently well. To get significant separations, the wire had to be so hot as to destabilize the gas.

Peierls was a theorist, so he embarked on a study of the theory of several methods of separating

FIGURE 2.1. Uranium hexafluoride (hex).
(Image by Bryan Derksen, Wikipedia)

uranium isotopes. His collaborator was Klaus Fuchs and in 1942 they produced a fundamental and certainly classified paper on the subject.[2] There is no doubt that Fuchs transmitted this paper to the Russians. The Soviet program at the time was still fairly modest. It became of the highest priority only later. On the first page of their paper there is a reference to a report by Dirac. P. A. M. Dirac was one of the greatest theoretical physicists of the twentieth century. In 1933 he shared the Nobel Prize in Physics with Erwin Schrödinger. Both men got the prize for work related to quantum theory. In 1928 Dirac had formulated the equation that bears his name, which is a unification of quantum theory and the theory of relativity. A product of this union was the prediction of antimatter. Indeed, in 1932 the American physicist Carl Anderson had observed the antielectron—the "positron." If you were to ask the average physicist what the Dirac equation is, you can be quite certain that the answer would be his 1928 masterpiece. But if you asked a centrifuge engineer the same question, you would get a quite different answer. It is worth noting that Dirac's early training was in engineering. If he had not gotten a fellowship to Cambridge, he might have ended up as an engineer—very likely one of the most brilliant engineers who ever lived. Dirac's centrifuge equation is a triumph of engineering physics.

What Dirac did was to ask, What is the maximum separation power that a centrifuge with a certain set of characteristics can have? These characteristics include such things as the length of the cylinder and its peripheral speed. In this work Dirac was not concerned with some actual centrifuge. It would have additional characteristics, such as the friction of its bearings or its stability against vibrations, which would prevent it from achieving this ideal. There was a precedent for this line of inquiry. In the nineteenth century the steam engine had been introduced. Early in that century the French engineer Sadi Carnot had asked how efficient such an engine could be. Again he was not considering some specific engine but the most general abstract engine.

To take an example, suppose there is a machine that can extract heat from a reservoir. It uses part of this heat energy to do some work. The rest it deposits in a reservoir at lower temperature. This is all we know and all we need to know to place an upper bound on its efficiency. It is proportional to the difference of temperatures of the two reservoirs. If ever we build a machine to carry out this task, it can never be more efficient than this. Dirac's analysis is in this general sort of spirit.

I will not give Dirac's derivation here. It is both remarkably simple and very subtle. But here are two of the results. The maximum separative power is proportional to the length of the cylinder. This is not too surprising, since for a given radius this determines the dimensions of the cylinder. What is surprising is the dependence on the peripheral speed, v_p. The separative power goes as v_p^4—peripheral speed to the fourth power. This means that if you double the speed, the maximum separative power goes up by a factor of 16. It turns out that real-world centrifuges have a dependence of only v_p^2. Nonetheless, there is a premium on maximizing the rotation speed. Faster is better. But Dirac did something else. He introduced a way of computing how much separative work is required to carry out some specified task, and his method has been used ever since. I will give examples later. We can think of it something like this. I have a certain sum of money. This is the separative work that an actual separation process can perform. I know how much a bottle of milk costs. This is the separative work for a task. I can then say how many bottles of milk I can buy with the money at hand. In the case of separation, if I have a task like producing a kilogram of uranium enriched to a certain percentage, I can, using Dirac's value function, determine how much separative work I will need.

I have a confession to make. When a few years ago I decided that I would like to learn about this subject, I looked at a definition of the value function. I nearly gave up there and then. It is not that it is so complicated mathematically. It just involves logarithms and a few simple algebraic polynomials, expressions like $(1-2x)\ln((1-x)/x)$. It is that I could not see, as the great nineteenth-century Scottish physicist James Clerk Maxwell would have said, "the go of it." If you want to see a derivation with "the go of it," you can look at the paper by Peierls and Fuchs I referred to earlier. I am instead going to focus on how the derivation is applied. In Figure 2.2 I have drawn a very schematic sketch of some

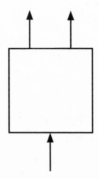

FIGURE 2.2. An abstract centrifuge diagram with the
feed, the product, and the tails shown by arrows.

separation facility. Later I will give examples in terms
of cascades of centrifuges.

I want you to notice the three arrows. The bottom arrow represents the "feed." If we begin with
hex involving the mixture of isotopes that occur
naturally, it will contain 0.711 percent uranium 235
and over 99 percent uranium 238. Incidentally, we
may wonder why there is such a big difference in
concentration, since all the isotopes of these heavy
elements were created at the same time in presumably similar amounts in supernova explosions. The
reason is radioactivity. These isotopes are radioactive. One usually gives their half-lives—the time it

takes for half of any sample to decay. For uranium 238 it is about 4.5 billion years, while for uranium 235 it is only 704 million years. Uranium 235 has decayed away faster. We can be grateful. If 235 were as plentiful as 238, nearly every country with any kind of industrial base would have nuclear weapons. Back to our abstract separation process.

The top two arrows represent the outward flow. One portion of the flow is the "product"—a certain number of kilograms that have been enriched to whatever degree we have decided we need. The other arrow represents what in this business is usually called the "tails." This is leftover hex from which some of the uranium 235 has been removed. The uranium is "depleted." The operator of the separation facility can choose what the concentration of uranium 235 should be in the tails. The smaller the concentration chosen in the tails, the more separative work is required for a given concentration in the product. The image that is sometimes given of this is squeezing oranges. If we want a certain amount of orange juice, we can choose fewer oranges and squeeze harder or we can choose more oranges and squeeze less. The choice might depend on the cost of oranges and our willingness to squeeze. Generally tails concentration is chosen to be between 0.2 and 0.3 percent as opposed the 0.7 percent concentration in natural uranium.

The separative work for a given task is stated in what are called separative work units—SWU—pronounced "swoo." You can find calculators on the web that will enable you to set a task and the tails and find the SWU.[3] Here are two examples. Suppose in both cases you start with natural uranium hex, which has a 0.71 percent concentration of U-235, and you set the tails concentration to 0.25 percent. In one case you want to end up with 1 kilogram of hex with a 90 percent concentration of uranium 235. This is called "highly enriched uranium"—HEU. In the other case you want to end up with 1 kilogram of 3.5 percent enriched hex—"low enriched uranium"—LEU. HEU is suitable for bombs, and LEU is suitable for reactor fuel. The division point between the two is set at 20 percent enrichment. Using the calculation table you will find that the first task using a 0.25 percent waste requires about 208 SWU, and the second requires about 13 SWU. If you begin with 20 percent enriched uranium and enrich it to HEU, it takes about 19 SWU. The question is, are these big or small numbers? And why do they get so much smaller when you begin with 20 percent enriched? It is the first step that requires so much separative work. The "breakout" to highly enriched uranium is much less costly in separative work.

We immediately get a sense of what the numbers mean practically—are they large or small—when we learn that a very fast centrifuge can produce about 5 SWU a year! This tells us that the production of HEU or LEU on any reasonable time scale is going to require a great many centrifuges working together in what are known as "cascades." Designing these cascades seems to me to be both a science and an art. There are two basic linkages, "parallel" and "serial," and they serve different purposes. Figure 2.3 is a crude representation of a parallel connection. Each centrifuge is being fed with hex, but the product is the same for both. This is opposed to the serial connection shown in Figure 2.4. In that figure, the product of one centrifuge is fed into the next to enhance the separation.

Figure 2.5 is how a typical cascade might look. In an actual cascade the number of centrifuges diminishes as we approach the "product." This is basically what accounts for the somewhat paradoxical SWU numbers I quoted before. It is less costly in SWU, and by substantial amounts, as we take our initial condition farther up the enrichment ladder.

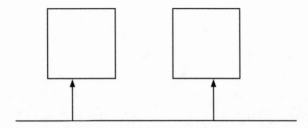

FIGURE 2.3. A parallel centrifuge connection.

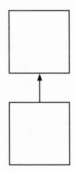

FIGURE 2.4. Two centrifuges connected in a series.

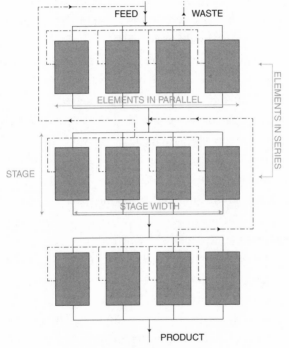

FIGURE 2.5. A model cascade.
(Courtesy of Ivanka Barzashka, Federation of American Scientists)

I will finish this chapter by noting an additional feature of these designs. One arranges the "plumbing" so that the tails from any centrifuge are recycled. Diagrams of this cascade plumbing I find resemble a Mondrian drawing.

It is important to note that what this configuration produces is enriched hex. What one wants is enriched uranium. We may ask how much uranium 235 by mass is contained in a kilogram of hex that has been enriched to 3.5 percent. To carry out this calculation we must remember that the atomic weight of fluorine is about 19 and that there are 6 fluorines to each uranium atom in hex. So the mass of U-235 in such a kilogram is about $.035 \times 235 / (0.65 \times 238 + 0.035 \times 235 + 6 \times 19) \approx 23$ grams. Or in words, for each kilogram of 3.5 percent enriched hex you might expect to recover at most 0.023 kilograms of uranium 235. The uranium in hex must be turned into a metal, which can be done with very little loss. This is done with a reduction process that strips the hex of its fluorine, producing, for example, uranium oxide, a powdery combination that can be made into pellets for reactor fuel. But these pellets will also have only a 3.5 percent enrichment—we do not extract the U-235—which means that although they are useful as reactor fuel, they cannot be used to make a bomb.

I think that I have said enough for you to see that designing these cascades is not a trivial matter. Neither is running them.

In Chapter 3 I will return to the centrifuges, which are the basic components. But let me finish here with an another estimate. The power reactor at Bushehr, which produces about 1,000 megawatts of electric power, requires an annual fuel consumption of about 28.6 tonnes (1 tonne = 1,000 kilograms) of 3.5 percent enriched uranium. To get this from the 3.5 percent enriched hex, you need somewhat more than 30 tonnes of hex. But in the conversion to the uranium oxide pellets, perhaps a percent of the uranium is lost. So let us say you need annually 33 tonnes of 3.5 percent enriched hex. But it takes about 5 SWU per kilogram to enrich from natural uranium to 3.5 percent. So we need about 165,000 SWU per year to supply this hex. The standard centrifuges in Iran produce about 1 SWU per year. There are less than 20,000 of them. On the other hand, the Russian capacity is about 20 million SWU per year. That is why for now and in any foreseeable future the Bushehr reactor will use imported fuel from Russia.

3

Unintended Consequences

ON MAY 7, 2008, Gernot Zippe died in Bad Tölz, Germany. He was ninety. I could find only one obituary for him. It was in the URENCO trade magazine—URENCO being the European consortium that enriches uranium commercially with centrifuges. In fact, even though Zippe and I had been in occasional communication by phone and email for a few years before his death, I did not learn about it until some months after the fact. Through no fault on Zippe's part, his centrifuge has been responsible for a very substantial portion of the proliferation of nuclear weapons to places like Iran and North Korea. To blame Zippe would be like blaming Einstein for nuclear weapons because he invented the equation $E = mc^2$. Zippe once remarked, "With a kitchen knife you can peel a potato or kill your neighbor, it's up to governments to use the centrifuge for the benefit of mankind."

After it became clear that Zippe's English was better than my German, all our communications were in English. While he answered many of my questions about the centrifuge, I never could get him to tell me much about his life. But I learned a good deal about it later. He was born in 1917 in Varnsdorf in what was then Austria. The family moved to Vienna, and he got his PhD at the University of Vienna, basically in engineering, in 1939. While he was getting his higher education he had piloted airplanes and gliders, and he wanted to be a fighter pilot in the Luftwaffe. But he was made a flight instructor for the duration of the war. He was still flying planes when he was eighty. During the war he was captured by the Russians near Prague along with someone he knew from Vienna, who explained to the Russians that Zippe had a PhD in engineering. To understand the next step, we must introduce the flamboyant character of Manfred von Ardenne.

Ardenne was only too willing to talk about himself. He even wrote an autobiography. The problem was what to believe. In any event he was born in 1907 in Hamburg into a wealthy aristocratic family. By age fifteen he had received his first patent in electronics. By the end of his life he had some 600 patents in various fields. He never completed his formal education, and when in 1928 he came into his inheritance, he used it to establish a private laboratory on an estate in Berlin. When the war came, his laboratory, which by this time was financed by the German post office, began to devote itself to nuclear energy. I have never been able to learn exactly what Ardenne's intentions were. Was he thinking of a bomb? The official German program, which had people like Werner Heisenberg attached to it, seemed to look down on the work being done by Ardenne's group as an irrelevance. Actually in some ways Ardenne's people got at least as far, if not farther. Two things are worth mentioning.

In 1940 the Dutch-Austrian physicist Fritz Houtermans, who had the unfortunate distinction of having been imprisoned in Russia by the NKVD and then having been turned over to the Gestapo, found employment in Ardenne's laboratory. Houtermans soon proposed using element 94—which had secretly been named "plutonium" in the United States—as a fissile material. He was so worried about the implications of this that he tried to warn people he knew in the United States. The same plutonium suggestion was made at about the same time in the official program by Weizsäcker. Having read both proposals, I found Houtermans's much more sophisticated. Of more relevance to us was the effort of Ardenne's laboratory to separate the isotopes of uranium. Ardenne used an electromagnetic method much like the Calutron that was used at Oak Ridge to make the final separation of the isotopes. It is an industrialized version of a mass spectrograph. I have not been able to learn how much, if any, these Germans actually separated, but when the Russians occupied Berlin they went straight to Ardenne's laboratory. Ardenne had thoughtfully posted signs in Russian outside his laboratory saying that it was in fact a scientific laboratory. Ardenne, some assistants, and his equipment were taken, more or less voluntarily, to the Soviet Union. Houtermans was long gone.

By the time of this expatriation Stalin had decided that Russia must have an atomic bomb. The Allied monopoly was, he felt, a guarantee of the limited expansion of the Soviet Union into western Europe. One of the things that he did was to appoint the much-feared head of the secret police, Lavrentiy Beria, to head the program. On one occasion, it is said, Beria went to Stalin to complain that some of the physicists seemed to be straying off the ideological reservation. Stalin said to him, "Leave my physicists alone, we can always shoot them later." Ardenne was taken to visit Beria in Moscow and told that his job was to help the Soviet Union build a bomb. Ardenne persuaded Beria that the activities of whatever laboratory was created for him should be limited to the separation of uranium isotopes. To this end an "Institute A" was established in Sukhumi in Georgia on the Black Sea. Ardenne devoted himself to electromagnetic separation as his group had been doing during the war. In a nearby town an "institute" was established, directed by the 1925 Nobel Prize–winning physicist Gustav Hertz. Hertz had Jewish ancestry, which prevented him from working in the German nuclear program during the war. He had been more or less concealed by the industrial firm Siemens. In his institute Hertz set about separating isotopes by diffusing uranium gas through tiny holes in membranes and taking advantage of the different rates at which the isotopes diffused. Both of these techniques were successfully used in our program at Oak Ridge.

The physicist Max Steenbeck had also been at Siemens during the war. When he was captured by the Russians, he was put in a concentration camp. He managed to write a letter to the NKVD explaining his scientific background. He too was sent to Sukhumi, but only after he went through a period of recuperation on a cream diet. Steenbeck devoted himself to centrifuge theory, although prior to this time he had never worked on centrifuges. He was soon joined by Zippe, who also had never worked on centrifuges. Zippe took charge of the experimental program. I asked Zippe what they had available to them. One thing was a 1940 Russian centrifuge—a bulky structure that weighed nearly a ton. To put such a thing in a cascade was unthinkable.

However, what was crucial is that they had available to them a 1938 article by Jesse Beams entitled "High Speed Centrifuging."[1] This article was almost a road map as to how to build such a centrifuge. Here are a few of the important points. After describing some of the difficulties of having the centrifuges operating in air—uncontrollable air currents are generated—Beams argued that the centrifuges should be operated under vacuum. Each centrifuge rotor would be enclosed in a vacuum chamber. From the references given in his paper it would seem that he had begun to implement this program. These rotors make contact by means of bearings. The bottom bearing is a single needle point that is constantly oiled. The top bearing is something else. Beams does not go into the early history.[2] However, at the beginning of the twentieth century the idea of using magnetic forces as mechanical bearings was already well known. It was a fairly obvious extrapolation to use them in centrifuges. The great advantage is that magnetic forces are almost frictionless.

Perhaps the most important matter Beams discussed was the stability of the centrifuge. In the centrifuge there is a cylindrical rotor that is rotating at the speed of sound or greater. If such an object should oscillate and tip over, the result would not be pretty. I was told that Beams was asked to move his laboratory to the periphery of the University of Virginia campus after a few too many such events. Toward the end of the nineteenth century, a French engineer named Gustave de Laval had the same issue. He had been trying to improve the centrifuges that were being used to separate milk from cream by speeding them up. But at the time of his insight he was actually working on the shafts of steam turbines, which were made of steel and rotated. He had become frustrated with the frequent failures of the shafts. Out of desperation, it seems, he replaced the steel with bamboo-like rattan. Remarkably, it seemed to cure the problem. Unlike steel, rattan is flexible. When an oscillation begins to occur, the rattan bends and a gyroscopic force is created that self-balances the shaft. The theory of how this works was developed only much later, but it works. All materials have critical frequencies at which vibrations and oscillations are sharply enhanced. There are two types of centrifuge—subcritical and supercritical. In supercritical centrifuges the rotors pass through the region of critical frequencies. If the centrifuges were not made of flexible material, they could not do this without getting out of control.

Zippe and Steenbeck incorporated all these ideas and more. Figure 3.1 is a schematic diagram of what is commonly referred to as the "Zippe centrifuge," but it is not the one designed by Zippe. Whenever we discussed it, he always called it the "Russian centrifuge."

The heavy isotope flows to the bottom (waste) scoop while the lighter one flows to the top (product) scoop. This double flow is called "countercurrent" and was invented before the war by the American physicist Harold Urey. It serves a very important purpose. Without it the separation takes place radially outward from the center. The centrifugal force, which acts differentially on the molecules of different mass, increases with increased distance radially to the periphery of the rotor. But the countercurrents that are directed parallel to the rotor also separate isotopes. This is the same sort of diffusion

GAS CENTRIFUGE

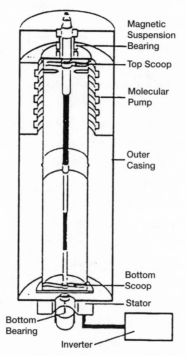

FIGURE 3.1. This is a schematic drawing of a gas centrifuge. The "scoops" collect the uranium hexafluoride, which travels in two currents moving in opposite directions. The top "bearing" is magnetic and hence friction free, while the bottom bearing is a needle which must be kept constantly oiled.

(Courtesy of David Albright; Illustration by Jandos Rothstein.)

mechanism that operates in the gas diffusion method of separating isotopes. Hence, there is both radial and axial separation, and the efficiency of the centrifuge is considerably improved. The product and tails are removed from the centrifuge by what are known as "scoops." At the top the rotor is held in place by a magnetic bearing that is more or less frictionless; at the bottom it is balanced on a steel needle. The German group put all these elements together better than anyone had done before. Their best centrifuges were made of aluminum tubes with a diameter of about 74 millimeters and a length of about 334 millimeters. This centrifuge, which was subcritical, had peripheral speeds of about 350 meters per second—faster than the speed of sound in air. One of the great advances they made was in power consumption. Their machine used 3 watts or less, which was better by a factor of 1,000 than any other centrifuge then in use. Later they constructed a centrifuge that was nearly 10 feet tall that had a somewhat smaller peripheral speed of 250 meters per second. Until they did it, no one believed that a centrifuge of this length could be made.

On March 1, 1948, on a trip to Moscow, Steenbeck was given an ultimatum. Either he would succeed in getting the centrifuge to separate uranium within a month or the project would be closed, which seemed to mean that the Germans would go back into detention. On March 21 they successfully separated the uranium isotopes. In 1950 Steenbeck had a meeting with Beria, who agreed that the Germans could sign contracts that had definite termination dates. In 1952 Zippe was transferred to Leningrad, where he worked for two years with some Russian colleagues. Then he was put in a transition camp in Kiev for another two years until in 1956, after some ten years in the Soviet Union, he was finally released.

After his release Zippe chose to resettle in West Germany. Ardenne and Steenbeck chose East Germany, where subsequently they had substantial careers. Zippe told me that he was not allowed to take any documents with him when he left Russia, but that this did not matter because he had the details of the centrifuge in his head. He was sure at the time that the reason the Russians had let him go was that they had lost interest in the centrifuge program. He learned later that this was wrong—they had a vast centrifuge program—so he was never sure why they let him go. Soon after he returned to Germany he learned that there was a conference on centrifuges in Holland. He attended and was surprised to learn that the Russian centrifuge was better than any of the others that were available. He spoke to, among other people, a Dutch centrifuge designer named Jacob Kistemaker. Kistemaker was so impressed that he changed his whole program so that it followed along the lines of the Russian centrifuge. Zippe became a consultant to Kistemaker. The delicate matter of patents was not raised by the Russians because they did not want to reveal their interest. As it happened, there were centrifuge programs in Germany, Holland, and England. The German company Degussa, which had an unsavory Nazi past, got into the business. During the Nazi years their subsidiary Auer had sold processed uranium metal to the German nuclear energy program for their reactor work. The most dangerous part of the work processing the metal was done by forced laborers from concentration camps. One of Auer's subsidiaries, Degesch, produced the Zyklon-B gas that was used in the German extermination camps. In 1939–1945 Degussa received at least five metric tons of gold to refine, taken from Jews who had been exterminated. Later Degussa tried to sell centrifuge technology to Saddam Hussein. Zippe became a consultant for the company.

In 1964 the Germans formed a state-owned enterprise to develop uranium separation technology.

In 1970 the company was privatized and signed an agreement known as the Treaty of Almelo. Almelo is a small town in the Dutch countryside where the Dutch centrifuge activities were taking place. There was a similar enterprise in England. URENCO, which conjoined these national activities, was founded in 1971. It is jointly owned by private entities in the three countries. Using descendants of the original Zippe centrifuge the company now produces about 15 million SWU per year. Zippe himself spent the years 1958–1960 at the University of Virginia with Beams designing improved centrifuges. He then returned to Germany, where he remained a consultant for industry. When I spoke to him he was living in retirement with his son. The stage is now set for the next actor, Abdul Qadeer Khan.

Khan was born in 1936 in Bhopal in what was then British India. His family was Muslim, so when India was partitioned in 1947, they left India for the newly formed state of Pakistan—first his brothers, then him. The family settled in Karachi, where his father became the headmaster of a school. After high school Khan enrolled at the Dayaram Jethamal Science College of Karachi, where he studied physics and mathematics. He obtained a BSc degree in

1960 from the University of Karachi, where he spe-
cialized in the physics of metallurgy. It is important
to keep in mind that Khan was a metallurgist and
not a nuclear physicist. After graduating, he worked
as an inspector of weights and measures in Karachi.
But it had always been his intention to complete his
studies abroad. In 1961 he resigned from his position and went to West Germany to study metallurgical engineering. He then went to Delft University
in Holland, where in 1967 he obtained a more advanced degree. He must also have learned Dutch,
something that turned out to be crucial for his
future activities. He later married a Dutch South
African woman. He completed his education at the
Catholic University in Leuven in Belgium where
the instruction was in Flemish—basically Dutch.
In 1971 he earned his PhD. By this time he knew
both German and Dutch as well as English. After
taking his degree he went to work for the Physical
Dynamics Research Laboratory (FDO) in Amster-
dam. It was a subcontractor to the Dutch branch of
URENCO. No one, at least in the Dutch branch,
paid much attention to security. Khan easily got access to the enrichment facilities in Almelo and was
soon asked to translate highly sensitive documents
from Germany involving centrifuge design—no

doubt variants of the Russian centrifuge—from German into Dutch. Security was so lax that Khan was allowed to take these documents home. In 1974 the Indians tested their first nuclear weapon and the Pakistanis became desperate to build a bomb. Khan was determined to help them.

On August 31, 2009, Khan, who had been in a very loose administrative detention in Pakistan, gave a television interview. He gave his version of the events.[3] The first question he was asked was, "What difficulties were you facing when you started the nuclear program of Pakistan?" This must refer only to the nuclear weapons program. When I was there in 1969 there was a nuclear program and indeed I was taken to see a nuclear reactor under construction. It had been supplied by our Atoms for Peace program, which also supplied Iran's first reactor, the so-called Tehran Research Reactor. Khan answers,

> Industrial infrastructure was nonexistent at that time in Pakistan. Immediately after the Indian nuclear tests in 1974, Zulfiqar Ali Bhutto [the prime minister] summoned a meeting of scientists in Multan [a city in the Punjab] to ask them to make a nuclear bomb. After the debacle of

East Pakistan in 1971, Bhutto was extremely
worried about Pakistan's security, as he knew
Pakistan had become very vulnerable. He
removed Usmani when the latter told him that
they could not go ahead with their plan of
acquiring a nuclear bomb because the basic
infrastructure was not there. [Ishrat Hussain
Usmani was a British-trained physicist who was
chairman of the Pakistan Atomic Energy
Commission from 1960 to 1972. I suspect that
he understood the physics of nuclear weapons,
which I doubt Khan ever did.] Usmani was not
wrong in his capacity. The Atomic Energy
Commission was the only relevant institution at
that time, but it lacked the required expertise.
India's nuclear test in 1974 created hysteria in
Pakistan. I was in Belgium in 1971, when the
Pakistan Army surrendered in then East Pakistan
and faced utmost humiliation. Hindus and Sikhs
were beating them with shoes, and their heads
were being shaved in the concentration camps.
I saw those scenes with horror. I was living in
Holland and working in a nuclear field. It was a
very useful field. At that time, there were only
three countries in the world that could enrich
uranium centrifugally. Though the United

States, France, China, and Russia were leaders in
uranium enrichment technology, they would use
the diffusion method instead of enriching it
centrifugally.

On the matter of Russia, as I have mentioned,
Khan was wrong. They had a vast secret program.
There was in the wartime Manhattan Project a
centrifuge program under Beams that was abandoned. The next question Khan was asked was,
"Who contacted you from Pakistan for this purpose?" He answered,

> I was not contacted from Pakistan. After the
> Indian test in 1974, I thought I must approach
> Bhutto and tell him about my capability of
> making a bomb. Though it was a very rare
> technology, I had firsthand experience of that
> technology and I knew how it worked. I wrote a
> letter to Bhutto in September 1974, telling him
> that I had the required expertise. Bhutto's
> response was very encouraging, and he wrote me
> back after two weeks, asking me to return to
> Pakistan. I came to Pakistan in December 1974
> to meet Bhutto. I briefed Munir Ahmer Khan
> [the nuclear engineer who succeeded Usmani
> as the head of the Pakistan Atomic Energy

Commission] and his team about this technology
and asked them to start creating the infrastructure before returning to Holland. I came again
[to Pakistan] in 1975. I used to come every year to
meet my family in Karachi. Bhutto asked me to
inspect the site in 1975 to check the progress if
there was any, but it was disappointing to see that
no progress had been made by that time. I told
Bhutto that I had to return to Holland, however
he insisted that I could not go and had to stay
here. I told him that I had a job there and I had
to go. I told him that I could only provide some
direction to local scientists. My daughters had
their education in Holland, and my wife had to
look after her elderly parents. I asked Bhutto to
give me some time to ponder and discuss with
my wife. When I told my wife that we were not
supposed to return to Holland she was shocked
and rejected the plan, saying a few moments later
that because of my credibility of not lying she felt
I could have done something for my country, I
said to her that I could claim without exaggera-
tion that no one could do it for Pakistan but me.
That is how she changed her mind and decided
to stay in Pakistan. She said we could visit
Holland as frequently as we liked.

If only Khan's wife had insisted on their returning to Holland for good, the world might be a much better place than it is.

In 1976 Khan left Europe under the pretense that he was going back to Pakistan on vacation, and then he did not return. He had with him the plans of the latest URENCO centrifuges. Khan was put in charge of the nascent Pakistani enrichment program, and soon gas centrifuges and their infrastructure were being built. Enrichment began in 1978, and by 1981 substantial quantities of highly enriched uranium were being produced. In May 1998 Pakistan announced that it had conducted five nuclear tests, all using highly enriched uranium produced with descendants of Zippe's Russian centrifuge.

There is little doubt that Khan's initial motives were patriotic. But what happened next reflects a mixture of greed and megalomania. In 1976 he was given his own enrichment facility—the Engineering Research Laboratories—located not far from Islamabad, the nation's capital. He soon came to think that all of its production, and indeed all bomb technology, was his personal property to sell. What is amazing is that there was some agreement that this was his right. General Mirza Asiam Beg, who was chief of staff of the Pakistan Army from 1988 to

1991, is quoted as saying, "Dr. A. Q. Khan and his scientists have given the country a credible deterrent for a paltry sum of money. What they have in their accounts is what I call gold dust—they have not taken the government's money. If a scientist is given 10 million dollars to get equipment, how would he do it? He will not carry the money in his bag. He will put the money in a foreign bank account in somebody's name. The money lies in the account for some time, and the markup that it fetches may probably have gone into his account. It is a fringe benefit." Khan did not need much encouragement like this, and soon he had set up a sort of supermarket in nuclear weapons technology.

Much of the operation was run out of Dubai. Customers like the Libyans could study a price list. The complete package, which the Libyans bought and which included all the elements needed to make the fissile material, as well as the bomb design itself, is said to have cost about $3 billion. The Libyans did not have the technical infrastructure to make use of much of the material. Moreover, in October 2003 a ship—the *BBC China*—that was carrying some nuclear technology from Malaysia, where Khan had some of his suppliers, was boarded when it stopped over in Italy. Colonel Gaddafi decided that this game was not worth the candle and used his nuclear booty as a trade for recognition by the United States. In December, CIA agents were dispatched to Tripoli to take possession of it. Just before the plane took off, the Libyans came to the airport and presented the agents with several envelopes containing plans for the nuclear device itself. It is said that some of them were in Chinese, a remnant of a deal that Khan had made to trade centrifuges for these plans.

In the late 1980s and early 1990s Khan dealt with the North Koreans and the Iranians. He made at least a dozen trips to Pyongyang, the capital of North Korea. Some of these were made in Pakistani army planes, indicating that the government was in collusion. Apparently there was a barter arrangement for nuclear technology. Khan traded bomb technology for North Korean rockets. It is not known what the North Koreans got, because so far they won't tell us. We can't find out from Khan either. In the television interview I have been quoting from, Khan refused to say anything about North Korea. But on Iran he was somewhat more forthcoming: "Iran was interested in acquiring this technology. Since Iran was an important Muslim country, we wished Iran to acquire this technology.

Western countries pressured us unfairly. If Iran succeeds in acquiring nuclear technology, we will be a strong bloc in the region to counter international pressure. Iran's nuclear capability will neutralize Israel's power. We had advised Iran to contact the suppliers and purchase equipment from them."

Needless to say, the "suppliers" were Khan's own company and the "equipment" was surplus taken from Pakistan's nuclear weapons program. In Chapter 4 we turn to Iran.

4

God the Merciful, the Compassionate

IN 1975 THE SHAH OF IRAN signed a deal
with the German company Kraftwerk Union AG to
build reactors some eleven miles from the city of
Bushehr on the coast of the Persian Gulf. Whatever
the Shah's intentions were, his action led Saddam
Hussein to begin his own reactor program with the
French. This was the Osirak reactor that the Israelis
destroyed in 1981. This reactor, unlike most at this
time, was to be fueled by highly enriched uranium,
which had not yet been loaded. The uranium was
not destroyed in the raid and enough was left to have
been used for a nuclear weapon.

On September 22, 1980, Iraq invaded Iran and
the Iranian attitude toward the nuclear program
began to change. The speaker of the parliament,
Hashemi Rafsanjani, began an effort to bring scientists back to Iran. In 1989 Supreme Leader Ruhollah

Khomeni died and Ali Hoseyni Khamenei became Supreme Leader, which he still is, and Rafsanjani became president. It was clear to Rafsanjani that without foreign help the Iranians were not going to be able to realize their nuclear ambitions. They made deals with China and North Korea, among others. Originally the Chinese sold the Iranians about two tons of uranium ore and plans for a uranium reprocessing plant. But in 1997 the Chinese backed off in response to American pressure. The Iranians did not report these purchases to the International Atomic Energy Agency (IAEA) and simply used this material covertly. The North Koreans sold the Iranians missiles and some technology for mining uranium. Beginning in the 1980s the Iranians had secret, but official, government dealings with Pakistan. These took a different turn when A. Q. Khan was brought into the picture. There were initial contacts in 1987 between the Iranians and Khan's representatives in Dubai. By 1993 the Iranians were offered a "menu." This included a sample "P1" centrifuge, the details of which I will describe later, along with plans for the more advanced P2 centrifuge and apparently the plans for setting up a cascade. There was a good deal of auxiliary equipment on order. The price list ranged up to hundreds of millions of dollars. It seems as if the Iranians bought the plans for making metallic uranium hemispheres, whose only use is in nuclear weapons. I think it is probable that they bought the Chinese design for such a weapon, which is what Khan had sold the Libyans. Because no Western intelligence officer has been allowed to interview Khan, we cannot be sure. Much of the stuff the Iranians bought was secondhand. Much later when the IAEA inspectors examined the P1 centrifuges, they found traces of highly enriched uranium, indicating that the Pakistanis had used them in their enrichment program. The Iranians seem to have bought a number of things that were junk—castoffs—but they had enough material about centrifuges to start their own uranium enrichment program. They set up a fake company called the Kalaye Electric Company, which they claimed was a clock factory, in a suburb of Tehran. Here they made P1 centrifuges and the hex to feed them with. The hex was first fed into the centrifuges in 1999, and in 2000 the Iranians set up their production facility in Natanz in central Iran.

By this time Gholam Reza Aghazadeh was in charge of the Iranian program. He was born in Khoy,

Iran, in 1949, earned a BSc degree in accounting and computer engineering from the University of Tehran, came to the United States for further studies, but in 1978 returned to Iran, where he participated in the revolution. He became director of a newspaper that was run by Mir-Hossein Mousavi. When Mousavi became foreign minister in the new government, Aghazadeh became his deputy for economic affairs, and when Mousavi became prime minister, Aghazadeh was put in charge of the country's oil dealings, becoming minister of petroleum in 1985. In 1997 he became the vice president of Iran in charge of atomic energy, a post he held until July 2009 when he suddenly resigned after Mousavi lost the Iranian presidential election.

On April 12, 2006, Aghazadeh gave a remarkable interview on Iranian television. This interview, from which I am going to quote extensively, gives, I think, a real insight into the Iranian nuclear program. It is especially interesting because of who Aghazadeh is and who his associates are.[1] I will first consider some of his comments on the spirit of the program, and then I will give a more technical analysis of what he reveals about it. When asked about what sort of thing went into carrying out the program, Aghazadeh said:

I will give you an example. In the preliminary stages of the work, we noticed that our machines broke down frequently. We couldn't discover the cause, since we didn't have any scientific sources or books to refer to. After great efforts we discovered that our experts didn't wear fabric gloves during the assembly phase. We found out that when you assemble the parts with bare hands, germs are transferred to the machinery from the smallest amount of sweat which comes off the hands.

[Because of the high speed of the rotors] this little amount of germs is enough to trouble and destroy the machine. When we say a machine is destroyed we mean that it turns into powder.

He begins, "The centrifuge used by us is about 1.80 centimeters high." Aghazadeh surely means meters here. His reference is to the Iranian version of the Pakistani P1 centrifuge. There are some odd technical statements in the interview, which might reflect the fact that Aghazadeh is not a physicist. With meters this conforms to the known dimensions of a P1 centrifuge. This centrifuge is much taller than the carbon-fiber rotor shown previously. That rotor can reach speeds of some 600 meters per second, as compared to the 350 meters per second of the P1.

That is why it can be shorter. Remember the Dirac equation for the maximum separative work, which shows that a shorter rotor can be compensated for by higher peripheral speeds. This centrifuge is actually a descendant of the successor to the newer P2, which, instead of having a rotor made of aluminum like the P1, was supposed to have one made of "maraging steel." This is a non-carbon-based form of steel that is tougher and more malleable than ordinary steel. The P2 is about half the length of the P1. The decrease in length provides a great advantage in the ease of manufacture. The longer centrifuges are made of shorter pieces that are attached to each other by what are called "bellows." These flexible attachments—first introduced in the Russian centrifuges, one of Steenbeck's ideas—are very difficult to manufacture. In fact, the Iranians were not able to make the maraging steel P2 bellows—a single one for each machine—which is why they switched to carbon-fiber rotors. The carbon-fiber centrifuges, known as the IR-2, are sufficiently short to require no bellows; they can be made in one piece. Their peripheral speeds are as much as 700 meters per second, twice that of the P1.

Now back to Aghazadeh. He continues, "Every centrifuge machine has 94 parts. [I do not know how he divides things up to get this exact number.] All these parts are sensitive. Twenty parts are high-tech. These pieces have manufacture plans which are highly complicated. Some centrifuge parts have 15 manufacture stages. They are high precision. All the measures in these parts are less than a micron. It is in millimeters in some ordinary parts but it is under a micron in other parts. The manufacture technology for these parts should have been created in Iran. This is one of our greatest successes."

The interviewer interrupts. "You mean every centrifuge has 94 parts." Aghazadeh responds:

> The machines differ with each other. The ones we use have 94 parts. And there are accessories such as the tap on top of the machine, etc. All that is being manufactured in Iran now. When these parts are prepared, the next stage is the assembly of the machine. That is a very sensitive and important thing. Particularly the assembly equipment. We cannot access assembly equipment in any part of the world. Our experts have designed these machines so that they would assemble the centrifuge machine. Those who have visited this equipment, found it very interesting that this modern equipment was being made here.

So much for the success of boycotts. Aghazadeh continues.

> When the machine parts are made, they should be tested for rotation. This is called balancing. Then the machine begins to rotate and raise this thing [*sic*]. At the final stage, a machine should reach 64,000 rpm. In terms of linear speed it is 350 meters per second. When I was on board a plane, I was comparing it to the speed of a plane. Normally planes travel seven meters a second. [Here is another odd slip. Jets fly at about 225 meters a second.] So, the raw material, technology, everything is at a high level. After rotation starts, the stage of injecting gas starts. When the gas enters the machine, it has an extraordinary combination. This is very sensitive in terms of pressure and settings. After two years of voluntary suspension [when they gave up working on nuclear weapons], which ended 2.5 months ago, we had a hard job to do. There is a huge complex which has been shown yesterday and today. We did all this during these 2.5 months. There is a stage when the machine is assembled, then it comes on the platform and becomes part of a chain and then a vacuum should be created. This

is easy in words but it is too complicated in practice.

One has only the translation. From what follows I think what is meant here is "very" complicated. Aghazadeh continues, "A vacuum system should be created. It is easy to talk about vacuum. But creating a vacuum is very complicated. You should then stabilize the system and inject gas to do the work."

Estimates that I have seen are that it takes a day or more for the injected gas to reach equilibrium before the separation begins.

Aghazadeh continues:

> Because the work was done for the first time, in order to succeed we primarily set up four single machines, then a chain of 10, followed by a chain of 20 and finally a chain of 164. I will explain about the philosophy of the chain of 164.
>
> Our first test on single machines was very successful. Then we worked on 10 and 20 machines. While working on 20 machines, we reached a very good degree of enrichment and this helped the scientists to become confident of the result. So we assembled the whole chain and as you are aware the enrichment was carried out.

First, why did we use a chain of 164? As a power plant fuel we need between 3 to 5 percent enrichment. . . . For example, we increase the percentage of isotope 235 from 0.7% to 3.5%. Naturally speaking, isotope 238 will reach 96.5 percent enrichment. [He means that the percentage of U-238 will reach 96.5 percent. I do not know why he uses the phrase "naturally speaking" unless this is some quirk of the translation.] In a chain of 10 this will create more than 1 percent enrichment. In a chain of 20, the enrichment will be close to 2 percent.

A number of machines must be designed and configured. When you inject the gas, your output will be between 3 to 5 [percent enriched]. Therefore the optimum number of machines that can give you this enrichment is 164. [These remarks are incomprehensible to me. The question is, how many SWU does the cascade produce?] When you configure this chain of 164 and install its auxiliary system, then there is a huge gas injection system. When the gas enters the system, it has a product and a residue [tails]. These are highly complicated systems. The most important thing is that these machines with this speed and with this technological quality, which is not visible, need highly advanced equipment that should identify all the parameters of the chain machines using computer systems for which we have to develop software and equipment and design systems. Probably this has been shown too. There is a very modern control room than handles this complex. The system receives these parameters from these machines every moment. It controls the rpm of the machine and its speed as well as pressure and temperature, vibration and its impact on other machines, the temperature of the cooling water, control of the electricity. All these parameters should be controlled round the clock. And there is the gas injection. When you inject the gas, you do not see it in the machine. We have to control these with automatic systems.

Aghazadeh explains:

After starting to inject gas into this chain—first let me tell you that there are various kinds of injections. It is not as if you can start the injection at the end of the 164-chain link and expect it to move all the way through to the other end. There are 15 steps in the process. Starting in phase 10 gas will be injected into 24 units. It is enriched to a small degree at this stage. The product will be fed to the next phase. The residue from this process will be fed to the next phase. This is a complex process. When a country is able to obtain such enrichment levels, then all that is left will be to increase the numbers—meaning another 164 units will be built in order to increase the production capacity.

Aghazadeh says something interesting about the uranium hexafluoride they used.

One of the characteristics of uranium is that when it is mixed with fluoride, which is a very poisonous gas, it turns into gas at 50 or 60 degrees of heat. [These are degrees centigrade. Hex turns from a solid into a gas at 137 degrees Fahrenheit.] It turns from powder into gas. The UF6 capsules that are produced in Esfahan [the city in Iran noted for its blue mosques where there is a very substantial uranium transformation complex both in and out of hex] are in the form of powder in capsules. The capsules weigh ten tonnes each. [Remember that the density of uranium is about 19 grams per cubic centimeter.] . . . They are placed in a section and put under temperatures of up to 50 to 60 degrees. The UF6 turns into gas. With the centrifuge machines we inject the gas and use the centrifuge capability. Because the U235 uranium isotopes are lighter than the U238 isotopes, the former goes to the wall and the latter to the center. [This is of course totally wrong. The opposite is true.] Then the 235 is collected from the rotators. [All of this is in the form of hex.] And that is it. Once the UF6 is enriched it is returned and the fluoride is powder.

I will return at the end of this chapter to discuss some of Aghazadeh's explanations for all of this activity, but I want now to take advantage of the details he provides on the working of the cascades to draw some inferences about the Iranian centrifuges as they were at the time of the interview. I am not aware of any other source of such information. The Iranians will not allow their centrifuge scientists to be interviewed. I once read a physics paper written by a physicist at an Iranian university. There was an email address, so I emailed him to see if he could put me in touch with an Iranian centrifuge expert. I got a polite response which said that that was impossible.

I begin with the definitions of the essential parameters.[2] Let N be the total number of centrifuges. This is the product of the number of cascades and the number of centrifuges per cascade. Using N in this way contains an assumption about the efficiency of the cascades. It is that the total production is simply proportional to the number of centrifuges. Nothing is lost to "leaks." Let S be the number of SWU per year per machine. We are going to measure the time T in months, so we need the conversion factor. With these definitions, the number of SWU supplied in T months is $S \times T \times N/12$ SWU. We wish to compare this to the number of SWU needed to perform some task, such as producing M kilograms of hex, enriched to some percentage. If we know the number of SWU/kilogram needed—S_{kg}—then the total number of SWU needed is $S_{kg} \times M$ SWU. If we equate these, we have the equation

$$S \times T \times N / 12 \text{ months} = S_{kg}M.$$

Depending on what facts we are given, we can use this equation to solve for an unknown, such as S or T. As I shall now explain, Aghazadeh gives us enough information to solve for S, although, because his interrogator was not a physicist, and neither was Aghazadeh, this information is not quite as unambiguous as one might want. I will give my reading.

What is unambiguous is N. Aghazadeh considers one cascade with 164 centrifuges; $N = 164$. He says this produces 7 grams per hour of 3.5 percent enriched hex. We might ask what is the production of uranium 235, which is what really interests us in the long run. To find this, we shall multiply the 7 grams per hour by the mass fraction of 235—that is, $7(0.035 \times 235) / (0.965 \times 238 + 0.035 \times 238 + 19 \times 6)$ grams/hour = 0.16 grams/hour of uranium 235, a lot less. But there are 8,766 hours in a year, so in a year this one cascade produces 1.402 kilograms of uranium 235. But this is mixed with the 238, so a metal made out of it will still contain only 3.5 percent uranium 235. To find S we need $S_{kg.}$ This we can find from the SWU calculator. To do this we need the tails fraction. A reasonable reading of Aghazadeh gives this as 0.36 percent. There is no a priori argument for this number. It is a matter of how the operator wants to run the cascades. With this choice we find that it takes 111 SWU to produce 1 kilogram of uranium 235 at this 3.5 percent enrichment. We can now put things together. The total number of SWU needed for the year at this rate of production is 111×1.4 SWU/yr = 238 SWU/yr. So per centrifuge we must divide by 164 to give about 1 SWU/yr. It is interesting to compare this to the Dirac maximum, which is about 5 SWU/yr. This shows that there is a good deal of inefficiency in the actual machines. In the last section of the book I am going to present the latest figures. But I want to close this one with some of the things that Aghazadeh says about the context. I think his remarks are very enlightening.

Here is what he says about the people working on the centrifuges: "Our young scientists are working day and night. It is only appropriate that I should say the manager for the inspection stage is a distinguished graduate from one of our best universities in the country. He was once asked by an official who was inspecting the complex about this time. The manager said he had not been home for more than 2.5 months. There is a place where they sleep and they are working the rest of the time."

When I read this I could not help thinking of Los Alamos during the war. Of course the mesa was home for them, but they worked day and night. What did these young Iranians think they were doing? The people at Los Alamos knew what they were doing. They were in a race with the Germans to build the bomb. They had no way of knowing that the Germans were never really in the race. Many of the people, such as Bethe, Peierls, and Frisch, were German-Jewish refugees. They knew what it would mean if Hitler won the war. They knew what it had already meant to relatives left behind. And they knew how good the German scientific establishment was. Many of these Germans had been their teachers and colleagues. Working day and night was the least of it. But with these Iranians, what was the hurry?

Aghazadeh never really answers this question, which is never really asked. One thing he does say is how the nuclear energy projects affect the whole Iranian scientific enterprise:

> When visiting one of our centers, the former minister [he does not say of what] said an interesting point. He said that the country's advancement is not only a national pride but it can also promote the scientific level of universities. And that is exactly the case today. When you visit the UCF [the conversion facility in Esfahan where mined uranium is converted into hex] reactors and heavy water projects you see that there are so many different branches of science involved. I can almost certainly say that I have not seen any branch of the core sciences and engineering sciences which is not used in the enrichment process.

I will come back to this at the end of the book, but I want to close this chapter with another statement by Aghazadeh, this most sophisticated, apparently liberal individual who apparently had some American education. It makes it clear what we are dealing with.

There is a complete consensus among Iranians. This can be proved by watching the reaction of people who contacted me on different TV channels like Jim-e-Jam, and some of which you may remember yourself.

In my telephone conversations with Iranians residing abroad during the programs, it was clear although some may not even have liked me personally, when it comes to the nuclear issue they had very strong enthusiasm. This proves that there is a full consensus among Iranians on this issue inside and outside the country, among all walks of life and all political currents.

Someone told me, and I believe that it is true, that if there was a referendum the number of votes in favor of the nuclear issue would be even higher than the public votes for the Islamic Republic [during the first days of revolution]. That's how united people are on the very important nuclear issue.

PART II

Plutonium

5

Reactors

Glenn Seaborg had something in common with many second-generation Americans. His mother tongue was not English. In his case it was Swedish. This came in handy in 1951 on the occasion of his winning the Nobel Prize in Chemistry. He was able to address the King of Sweden in their common mother tongue. Seaborg was born in 1912 in Ishpeming, Michigan. In 1922 his mother decided that she had had enough of Michigan winters and moved the family to a suburb of Los Angeles. This was the first piece of good luck in the early parts of Seaborg's career. It was good luck because his public school in Watts had an outstanding chemistry teacher who changed Seaborg's set of goals. He now wanted to become a chemist. It was also good luck because the University of California at Los Angeles was within commuting distance and charged no tuition for California residents. It was the Depression, his father had lost his job, and there was no money for a college education, so free in-state tuition made college possible for him. Seaborg made enough of an impression that he was hired as a teaching assistant and received strong recommendations for graduate school at Berkeley, where he also became a teaching assistant. Here his good luck continued. One of the faculty members at Berkeley was Ernest Lawrence. He had been at Berkeley since 1928, and in 1931 he invented the cyclotron. Lawrence had a genius for designing scientific apparatus. Among other things he designed the Calutron—the industrial-size mass spectrograph that was used to separate uranium isotopes at Oak Ridge. Lawrence also had a genius for raising money, and even in the Depression years he managed to raise enough to make larger and larger particle accelerators. I think that this interested him more than actually using the machines to discover something. This was left to his Berkeley colleagues. One of them was Edwin McMillan.

McMillan, who was slightly older than Seaborg, had gotten his PhD from Princeton in 1932 and then had gone to Berkeley, where he worked with the newest version of Lawrence's cyclotron. When the discovery of fission was announced in 1939, McMillan began doing experiments on it. What he needed was an intense beam of neutrons. He produced these in the accelerator by accelerating the nuclei of "heavy hydrogen"—the isotope that has one neutron and one proton in its nucleus, as opposed to a single proton. The heavy hydrogen was made to collide with a beryllium target, and this produced the desired neutrons. He then tried a uranium target. McMillan studied the resulting fission fragments, but nothing especially interesting showed up. However, the unfissioned uranium did show something very interesting. To understand it, we have to note that unstable isotopes can emit three kinds of radiation. There is a terminology for them that goes back to the beginning of the twentieth century. What are known as alpha particles are the nuclei of helium, which some isotopes emit. What are known as beta particles are just ordinary electrons, and what are known as gamma rays are energetic quanta of electromagnetic radiation. What was at issue here, was that beta radiation appeared to be emanating from the unfissioned uranium.

When uranium 238 absorbs a neutron, it becomes uranium 239. This nucleus emits beta particles. As I discussed when I explained why the different uranium isotopes were found in various abundances,

each type of decay is characterized by a half-life—
the time it takes for half of any sample to decay. Mc-
Millan observed a beta decay with a 23-minute half-
life. This was not a surprise. It had already been
observed by Hahn, Meitner, and Strassmann and
was known to come from uranium 239. But there
was a second beta decay with a half-life of 2.3 days,
which had not been observed before. McMillan
conjectured that this came from a previously undis-
covered element—a long-sought and elusive transuranic. Uranium has the atomic number 92—the
number of protons in its nucleus—so that this new
element would have the atomic number 93. McMillan was not a chemist, so he was not able to do a
chemical analysis. He asked a colleague who was
and who got it wrong, claiming that the new decay
was just another fission fragment. But Lawrence had
managed to construct an even more powerful accelerator than the one McMillan had used, and Mc-
Millan repeated his experiment, this time with a colleague named Philip Abelson. This time he got the
chemistry right, leaving no doubt that a transuranic
had been discovered. They wrote a short paper,
which did not give a name to the new element. Mc-
Millan had thought of one. The planet Neptune is
the next one after Uranus, so he named the element

"neptunium." This was not revealed until August 1945 with the publication of the so-called Smyth Report, which described the wartime nuclear program.[1] Having made his discovery, for which he was awarded the Nobel Prize in the same year as Seaborg, McMillan left for MIT in 1940 to work on radar. Before he left he had briefly collaborated with Seaborg, who had become interested in radiochemistry, and he gave Seaborg permission to continue the work. Together they had discovered evidence for an alpha particle decay that also did not seem to come from any known element. They conjectured that it came from the next element in the periodic table, 94, which Seaborg named plutonium—Pluto being the planet beyond Neptune. Seaborg and some new collaborators created a new, shorter-lived isotope of plutonium—plutonium 238—which clearly showed the alpha decay. This occurred in February 1941. The discovery and the name were not disclosed until the publication of the Smyth report.

We can summarize what we have learned so far about the transuranics in the sequence of reactions given below. Here *U* stands for uranium, *Np* for neptunium, *Pu* for plutonium, and *n* for the neutron. We follow the convention that the atomic number is the subscript on the left and the atomic weight is

the superscript on the right. The numbers below
the arrows are the half-lives.

$$_{92}U^{238} + n \rightarrow {}_{92}U^{239} \rightarrow \frac{\beta + {}_{93}Np^{239}}{23.5\,\text{min}} \rightarrow \frac{\beta + {}_{94}Pu^{239}}{2.33\,\text{days}}$$

In words, a uranium 238 nucleus absorbs a neutron,
becoming uranium 239. This nucleus is unstable
and beta-decays into neptunium 239 with a half-life
of 23.5 minutes. The neptunium 239 beta-decays
into plutonium 239 with a half-life of 2.33 days. Plutonium 239 is also unstable with a half-life of 24,110
years. That shows why we don't find plutonium 239
in the Earth's crust and also why it is stable enough
to become the stuff of atomic bombs.

The accelerator experiments produced minuscule
amounts of plutonium 239—micrograms. The radiation detectors had to have been extremely sensitive
to have picked up the radiation from such minute
amounts of material. But Seaborg and his collaborators were able to separate a tiny bit (micrograms) of
plutonium from the rest so that they could analyze
its chemistry. They manufactured about a half a mi-
crogram of plutonium 239. They wanted to see if
Bohr's argument about the fissile nature of uranium
235 applied to plutonium 239 as well. They did an
experiment in March, about a month after the first

discovery, that showed that indeed plutonium 239 was fissile, in fact more fissile than uranium 235. This meant that you now had a second road to making a nuclear weapon. Transuranics had gone to war.

After Pearl Harbor the Nobel Prize–winning physicist Arthur Compton created a branch of what became part of the Manhattan Project in Chicago. He called it the "Metallurgical Laboratory" to disguise its intentions. In April 1942 Seaborg arrived at the Met Lab with a single colleague with the instruction to set up a chemistry laboratory in what had been a student lab facility. At the time the laboratory was set up, very little was known about the physics and chemistry of plutonium. Not enough had been produced so that any of it was visible to the naked eye. Apart from the fact that it was fissile, there was the apparent advantage that it could be separated out by chemical means from whatever matrix it was produced in. You did not need the very difficult methods of isotope separation. But the chemistry involved was far from trivial. In his autobiography *Adventures in the Atomic Age: From Watts to Washington*, Seaborg explains why:

> The chemistry group's challenge was to come up with a process by which we could separate the

plutonium from all the material in the aftermath of the chain reaction. The process would have to work on a large scale. The plutonium would be present in a concentration of about 250 parts per million. That meant that there would be about a half pound of plutonium in each ton of irradiated uranium. The uranium would also contain a large selection of intensely radioactive fission products. So our challenge was to find a way to separate relatively small amounts of plutonium from tons of material so intensely radioactive that no one could come near: the separation [of the plutonium from the uranium] would have to be performed by remote control behind several feet of concrete. There could be no breakdowns requiring repairs because the radioactivity would keep anyone from approaching the apparatus once it started operating.

We would have to develop this process for an element that now [in 1942] existed in such minute amounts that no one had ever seen it. All our knowledge was based on secondary evidence of tracer chemistry—measurements of radioactivity and deduced reactions. Tracer chemistry was itself relatively new, deductions based on it were often subject to doubt.[2]

As I mentioned earlier, the Germans—Houtermans and von Weizsäcker—suggested using element 94, plutonium, in nuclear weapons. Weizsäcker, and through him Heisenberg, were persuaded that plutonium was the royal road. There was no isotope separation involved—"merely" chemistry. They had never created any plutonium, so they had no idea how difficult this chemistry is. We must also keep this in mind when we discuss the potential use of plutonium by the Iranians. As I will point out, they will have the means to create plutonium in its uranium matrix. But separating it is another matter.

Seaborg had to create a group to work with him. At the time he was very young and relatively unknown, and for security reasons he could not tell potential recruits beforehand what exactly they would be working on. The people he did recruit were not allowed to use the name "plutonium." They called it "49," trying to disguise the fact that it was element 94 in the periodic table. In the end all of this turned out to be futile because spies like Klaus Fuchs revealed everything. Nearly all the plutonium that Seaborg's group used initially had been manufactured in a cyclotron at Washington University in St. Louis that ran 24 hours a day for a year. By the summer of

1943 they had milligrams, enough to see it through a microscope. It is strange that the Germans, who had access to a cyclotron in Paris and who understood the potential of plutonium, never used the cyclotron to try to make any, although it was clear to everyone that using cyclotrons to make plutonium had very limited possibilities. I will next discuss nuclear reactors where plutonium is really made, but I note here that by 1943 Seaborg and his group had found a method for doing the separation. No one knew why it worked, but it was adopted by the people at the Dupont Corporation who actually did the separation.

A nuclear reactor is basically a device that utilizes the energy produced in controlled fission chain reactions. In the first instance this energy is manifested in the kinetic energy of the fission fragments. This energy is converted into heat, which can run, for example, turbines that produce electricity. There are three basic components connected with the nuclear part of this. First there must be fuel elements that contain the uranium that is going to be fissioned. Reactor fuel is very largely uranium, although plutonium can play a minor role. I have already mentioned that one of the quantum mechanical manifestations of the fission reaction is that its rate

FIGURE 5.1. Fuel pellets.
(United States Department of Energy)

is increased as the speed of the neutrons that initiate it is decreased. I have also mentioned that the neutrons that are created in fission move very rapidly—at speeds a tenth that of light. We want to enhance the fission rate. This is done by slowing the neutrons down to a speed of a couple of kilometers a second, by the action of what is called a "moderator." Examples will be furnished. Finally we need a "coolant" to keep the fuel elements at a low enough temperature that they don't melt.

To make the fuel element, hex is reduced to uranium dioxide powder, which is then made into a ceramic pellet such as the one shown in Figure 5.1. These pellets are bundled into long thin rods made

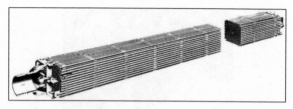

FIGURE 5.2. A "stack" of fuel rods from a nuclear
submarine reactor.
(Maritime Administration's Virtual Office of Acquisition)

of some metallic alloy. A collection of rods is called
a "stack," and stacks make up the core of the reactor. There can be as many as a couple hundred pellets in a fuel rod and a couple hundred fuel rods in
a stack and a hundred or more stacks in the core.
There are considerable variations among the types
of reactors.

As I have mentioned, fast neutrons are produced
by the fission that takes place in the core. These
must be slowed down, which is the function of the
moderator. The neutron and the proton have about
the same mass, so an ideal moderator would consist
of protons. Think billiard balls. When one billiard
ball collides with another, the first billiard ball can
lose a lot of its momentum to the second. Compare
this to a collision in which a billiard ball hits one of

the side cushions. Because of the discrepancy in masses the billiard ball can bounce off without losing much momentum. A natural source of protons is regular water, where there are two hydrogen atoms bound to one oxygen atom in each molecule and the nucleus of each hydrogen atom contains one proton. In the reactor business this is known as "light water." The reason for the "light" will become clear. But there is a catch. Protons can capture neutrons to produce a nucleus of heavy hydrogen—the "deuteron" with a proton and a neutron making up its nucleus. To conserve energy and momentum, a gamma ray is also produced. In symbols: $n + p \rightarrow D + \gamma$. When this happens the captured neutron is lost to the fission process. Nonetheless, many reactors use a light-water moderator. They make up for the neutron loss by using enriched uranium to enhance the fission possibilities. In the beginning of the reactor business several reactors used weapons-grade uranium enriched to more than 90 percent. This was the enriched uranium that was then available—a legacy of the war. Now reactors generally use low enriched uranium, LEU, with enrichments that go up to 20 percent. This is to avoid the burdensome restrictions that go with using weapons-grade uranium.

Some American nuclear submarines use 90 percent enriched uranium in their reactors to optimize the power.

Heavy water—water in which ordinary hydrogen is replaced by "heavy hydrogen," hydrogen in which the proton is replaced by a deuteron—used as a moderator is relatively free of the neutron capture drawback. It is true that deuterons can absorb neutrons to make "superheavy" hydrogen whose nucleus—the "triton"—consists of two neutrons and a proton. This is the reaction $n + D \rightarrow T + \gamma$. But its rate is much smaller than that of neutron capture by protons. The deuteron is about twice as massive as the proton, but it is still close enough in mass to the neutron that it is able to have the neutron's momentum transferred to it effectively. During the war the Germans tried to make heavy-water reactors but never could get enough heavy water to make it work.

The first successful heavy-water reactors were made in Canada in the late 1960s and early 1970s. These CANDUs—CANada Deuterium Uranium reactors—are "pressurized" water reactors. The heavy water is kept under high pressure to raise its boiling point and to keep steam from forming. This heavy water exchanges heat with a separate light-water coolant that does turn into the steam that runs electric turbines and such. When the Canadians began planning the CANDU, they lacked facilities for enriching uranium. But they realized that with the heavy-water moderator they could use unenriched uranium. They also made a very clever design in which individual fuel bundles can be removed from the core at the same time that fresh ones are being inserted. This means that the reactor does not have to be shut down while refueling takes place, an obvious advantage if the reactor is being used to generate electricity.

A drawback with the CANDU is the plutonium proliferation risk. I will go into this in more detail after I explain how plutonium is used in nuclear weapons. But here let me say that if the uranium fuel is left in the core too long, then undesirable isotopes of plutonium can be created. The fact that natural uranium is used is an advantage in plutonium production. Recall that plutonium is produced when uranium 238 absorbs a neutron, becoming uranium 239, which decays into neptunium 239, which in turn decays into plutonium 239. Thus, the less enriched the fuel is—the lower the percentage of uranium 235—the better it is for plutonium production. Natural uranium—uranium from a mine—with less than a percent by weight of uranium 235 is about as good as one gets. For these reasons, when one sees that a country like Iran is constructing a heavy-water reactor, one must pay some attention.

The first reactor ever constructed went "critical"—produced self-sustaining chain reactions—on Wednesday, December 2, 1942. The site was a squash court under Stagg Field—the football field of the University of Chicago. The architectural genius of this enterprise was Enrico Fermi, although the Hungarian-born physicist Leo Szilard made some important contributions. Among twentieth-century physicists Fermi was unique. He was probably the last physicist to understand, and to have made contributions to, every branch of physics. To give some idea, in the early 1930s Fermi wrote a basic review article on quantum electrodynamics—the unification of electricity and magnetism with quantum theory. He used ideas from this review to create the first theory of beta decay. At about the same time he invented the field of slow-neutron physics, from which the whole idea of moderators has its origins. As an experimental physicist he could be very hands-on, functioning like a practical engineer. Contrast this with Werner Heisenberg, who was the leader of the German reactor program. Heisenberg was one of the greatest theoretical physicists of the twentieth century. His papers on the theory of reactor design are very impressive. But like most theorists he liked to simplify. He liked models that were amenable to calculation. Unfortunately they did not work, and the Germans never were able to make a reactor despite Herculean efforts.

The moderator that Fermi chose for the reactor was graphite—basically carbon. One of Szilard's contributions had to do with the graphite. He found that manufacturers of graphite put in small amounts of boron for structural stability. But boron is an absorber of neutrons. Even one part in 500,000 would render graphite useless as a moderator, because too many neutrons would be absorbed and taken out of the fission cycle. Szilard pestered producers of graphite to make a boron-free product. Another very important contribution Szilard made was to the design of the reactor. Reactors come in two types, "homogeneous" and "heterogeneous." In a homogeneous reactor the moderator and the fuel are uniformly mixed. For example, in an "aqueous homogeneous reactor" they are both in liquid form and circulate as a single liquid. In a heterogeneous reactor the two are separated. Szilard argued that the design should be heterogeneous, and indeed most reactors follow this design.

The graphite in Fermi's reactor was machined into blocks. The idea was to make layers of these blocks that when completed would stack up like a child's version of a hemisphere made with toy blocks. Alternate layers had holes 3.75 inches in diameter drilled into the blocks; the slugs of uranium, which were about 3 inches in diameter, were put into these holes. In the end there were 45,000 blocks with 19,000 holes. One thing the reactor did not have was a coolant. There was no need. Once it reached critical Fermi intended to run it for only an extremely short time, during which its power would be about a half a watt—the wattage of a very, very dim lightbulb. There was a supply of ambient neutrons generated by the spontaneous fission of the uranium. The uranium occasionally split apart on its own accord. Once it got started, the reactor was controlled by the insertion of cadmium rods, which absorbed neutrons. During the test these rods were pulled out slowly, inch by inch—interrupted by a lunch break—until the reactor went critical, at which point they were reinserted and the experiment was over except for a celebratory glass of chianti. With this experiment the nuclear age really began.

The first thing that happened was that General Leslie Groves, who headed the Manhattan Project, had the Dupont company build a much larger reactor in Clinton, Tennessee. It too was a graphite-moderated reactor, but it was air-cooled and designed for 1,000 watts. It went critical in 1943 and produced over 300 grams of plutonium. The design of the X-10, as it was called, was relayed to the Soviet Union by a spy who has not been identified and became the basis of the first Russian reactor and hence of their plutonium program.

Given the success of the prototype, General Groves decided to go ahead with building the real plutonium production reactors, which would produce hundreds of kilograms. These reactors were going to be water-cooled, so the obvious thing was to locate them next to a river. Groves chose the Columbia River in Washington State, where it seems he had fished when he was a student at the University of Washington. Construction of the Hanford B-Reactor started in June 1943. This reactor used natural uranium and was graphite-moderated, and was designed to operate at 250 million watts. It began producing plutonium in September 1944, and the first delivery arrived at Los Alamos in January 1945. The plutonium then had to be reduced to a metal, which took until March. The metal sample weighed all of 520 milligrams.

But by this time enough had been learned about plutonium to change the program at Los Alamos in a fundamental way.

6

The Delta Phase

I HAD THE CHANCE to meet William "Willie" Zachariasen a few times when he came to visit his son Fred in Aspen, Colorado. Like his father, Fred was a physicist—he died in 1999—and we spent several summers together at the Aspen Center for Physics. I knew at the time that Willie must have been a distinguished physicist. He was a professor of physics at the University of Chicago and for a time chairman of the department, a department that included people like Fermi. I had no idea what field Willie was in and never thought to ask. He seemed to be a very amusing man with a wry sense of humor. I always thought he was like one of those sea captains who would get you through the storm while putting up with your frailties. I learned from Fred that Willie actually was the son of a Norwegian sea captain and that when he was a young man in Norway he used to row out to the islands in

Langesunfjord to look for crystals. Willie never lost his fascination for crystals and became one of the great crystallographers of the twentieth century. It was as a crystallographer that Willie played a vital role in the Manhattan Project.

In 1943 Willie joined the Met Lab. One of his tasks was to find the density of plutonium. Let me explain why in the bomb context this is so important. To understand this I will recapitulate my discussion of how to estimate the critical mass. We suppose here, as we did there, that we are dealing with a solid sphere of fissile material. Let us call the volume V and the density ρ. I noted in that discussion that there would be a critical size of the sphere V_c at which the number of neutrons that escaped from the surface of the sphere would be balanced by the number created in the fissions. The critical mass M_c would be $M_c = V_c \rho$. I also noted that $V_c \sim r_c^3$ where r_c is the critical radius of the sphere. I remarked that this critical radius must be of the order of the mean free path for fission, which is given by $1/n\sigma_f$ where σ_f is the fission cross section and n, in this instance, is the number of plutonium nuclei per cubic centimeter. But the mass density ρ—the number of grams of plutonium per cubic centimeter—is just n multiplied by the mass of a plutonium nucleus. So if we put things together, we see that $M_c \sim 1/\rho^3 \times \rho \sim 1/\rho^2$. Hence, if we have a way of, say, raising the density by a factor of 2, we lower the critical mass by a factor of 4. Density is a very important thing to measure, and this is what Willie tried to do.

He did not have much plutonium to work with. He was given 100 micrograms of powdered metallic plutonium. The metallic plutonium had a crystal structure, which meant that there was what was called a "unit cell."

Any sample of plutonium would consist of vast numbers of repetitions of the unit cell. Willie's plan of attack on measuring the density was to measure the size of the unit cell and hence its volume. Knowing the number of plutonium atoms in the cell, he could then find the mass per unit centimeter cubed—that is, the density. The assumption was that if there were impurities, their amount would be insignificant. Willie was not bothered by the fact that the sample was only milligrams. He had dealt with tiny samples before. But he was bothered by the results. They were inconsistent. He got 13 grams per cubic centimeter in one case and 15.5 in another, much too big a discrepancy to be accounted for by traces of impurities. Willie was just too good for this to be experimental error. Something was going on. Willie soon realized what it was—allotropes.

Allotropes are something we are all familiar with although we may not know the term. An excellent example is carbon. Carbon has eight allotropes, but two that you certainly know are diamonds and graphite. Both are composed of carbon, but their crystal structures are very different. This leads to very different physical properties. You may well give your sweetheart a diamond ring, which might make her happy, but if you gave her a lump of coal the message you would be sending is quite different. In the case of plutonium there are six allotropes, but in the first instance Willie found two. Now we would list all six by advancing the letters in the Greek alphabet as the temperature at which a given allotrope is stable increases. The allotrope of plutonium that is stable at room temperature is the alpha phase.

The alpha phase is much less symmetrical and has more plutonium atoms packed in, which is why it is denser. A solid made of alpha-phase plutonium acts more like a chalk than a metal. Its density, which is 19.6 grams per cubic centimeter, is substantially larger than the delta-phase density, which is 15.92 grams per cubic centimeter. This accounted for Willie's result. His samples had different allo-tropic mixtures. But there was a problem.

The alpha phase is stable up to a temperature of about 252 degrees Fahrenheit. But the delta phase is unstable below a temperature of about 603 degrees Fahrenheit. Below this temperature it morphs into the alpha phase. Thus, as it stands, the delta phase, which has all the desirable qualities of a metal, is useless in making a bomb. Enter Cyril Smith.

Smith, who was born in 1903 in Birmingham, England, got his PhD from MIT. As a metallurgist he got a job with the American Brass Company in Connecticut. When the war came he took a desk job in Washington with the War Metallurgy Committee. While attending a meeting in New York, he was approached by one of Seaborg's colleagues, Joseph Kennedy, who was going to New Mexico to head the chemistry department at about-to-be-created Los Alamos National Laboratory. Why Kennedy chose Smith is not clear, because Smith had not published much outside of his specialty—brass. Kennedy could not tell Smith much of anything about the Los Alamos lab, but there must have been some connection, because the next thing that happened was that Smith got a visit from Oppenheimer in

Washington. Oppenheimer also could not tell Smith much of anything, but Oppenheimer had a way of not saying much of anything which made it sound like it had very great importance. In any event, in March 1943 Smith became one of the first Los Alamos staff members.

It was decided not to try to do metallurgical work on plutonium until at least gram samples arrived at the laboratory, which occurred in March 1944. These samples could be reduced to a metal. It was soon confirmed that what Zachariasen had discovered about plutonium allotropes was correct. Smith had spent his career creating alloys of copper and zinc—brass—for various purposes, so it was natural for him to try to alloy the delta-phase plutonium with something else to stabilize it. There was no theory as to how to do this, and as far as I know there is none now. It was and is a matter of trial and error. The first alloy that worked was with aluminum. But there was a problem. When plutonium decays into alpha particles, these, when they interact with aluminum, produce neutrons. I will go into more detail as to why this is a problem for a nuclear weapon, but here let me say that these neutrons can cause the chain reaction to begin before one wants it to. Then Smith found that an alloy with gallium also stabilized the plutonium and did not have the undesirable neutron production. But no one was sure how long the gallium would maintain the stability. By now there was an urgency to finish the bomb, and it was decided to take a chance. The bomb that was tested at Alamogordo on July 16, 1945, and then dropped on Nagasaki on August 9 was an amalgam of delta-phase plutonium with 0.8 percent by weight of gallium. There is an additional benefit to this arrangement. After the chain reaction begins and energy is produced, some instability is produced in the delta phase, which to some extent morphs into the alpha phase. But this phase is more dense and hence, using the argument I gave before, the critical mass is smaller. What was critical for the delta phase is now supercritical for the alpha, which enhances the explosion.

Smith's successful work was done in the spring of 1945, but a year earlier something had occurred that threatened to completely rule out plutonium as a nuclear explosive. By that time gram quantities were arriving at Los Alamos from the Tennessee reactors. When the first batch arrived, something was observed that portended a disaster—spontaneous fission. As early as 1940 some Russian physicists had discovered that these heavy elements can fission spontaneously without the intervention of a neutron. So some spontaneous fission was expected with plutonium. The reason this matters is that this fission produces neutrons. If these neutrons are produced before a supercritical mass is assembled, there will be a predetonation—what in this business is called a "fizzle." There are fizzles and fizzles, so one must put this one in context. To make the dimensions of nuclear explosions comprehensible they are restated in terms of an equivalent amount of TNT. The bomb that destroyed Hiroshima had an equivalent power of about 15,000 tons of TNT, while the bomb that destroyed Nagasaki had an equivalence of over 20 kilotons. Hydrogen bombs produce an equivalence of millions of tons. When an atomic bomb produces, say, 500 tons equivalent rather than kilotons, this is classified as a fizzle. Keep in mind that the Ryder truck that Timothy McVeigh used in 1995 to destroy the Murrah Building in Oklahoma City contained only 2.5 tons of high explosive. A nuclear bomb fizzle could bring down much of a city.

It was anticipated that uranium 235 and plutonium 239 would both spontaneously fission, but that the rates would be small enough that this would not be an obstacle. The question was how many neutrons would be produced in what is called the "critical insertion time." This is the time it takes to assemble a supercritical mass. At Los Alamos the proposal was to do this assembly by using conventional explosives to fire one subcritical mass into another one—something that is called "gun assembly." The assembly time was the order of a tenth of a millisecond. So the question became how many spontaneously created neutrons would there be in, say, a millisecond. For uranium 235 there are about 1/100 spontaneous neutrons produced per second per kilogram. Thus in, say, a 50-kilogram mass, the number of spontaneous neutrons produced in a millisecond is negligible. This is also true of uranium 238, which is anyway less than 10 percent of the highly enriched uranium. For plutonium 239 there are about 17 neutrons per kilogram per second, so again in a millisecond the number is negligible. But when we come to plutonium 240 the story is entirely different. There are about 1.2 million neutrons per kilogram per second. Thus, in a millisecond a 10-kilogram sample of pure plutonium 240 will emit some 9,000 neutrons. In the reactor sample there might be a concentration of 1 percent plutonium 240. A few stray cosmic ray neutrons can start a chain reaction. So even this percentage of plutonium 240 nuclei can cause a premature chain reaction in the bomb. Where did this plutonium 240 come from and what could one do about it?

Where it came from was easy to answer. It came from the reactor. If the plutonium is not reprocessed within a few months, the plutonium 239 can absorb another neutron and become plutonium 240. That is why a reactor like the CANDU is a proliferation risk. It can be refueled at any time without taking it off line. Reactors can tolerate about a quarter of the produced plutonium being plutonium 240, but for a bomb you need about 7 percent or less. The plutonium from the Tennessee reactor already had too much plutonium 240 for a gun assembly device, and the Hanford reactors with their higher neutron densities would be even worse. What to do about it was not so easy to answer.

The first thought might be to try to separate the isotopes. But this seemed all but hopeless. It was hard enough to separate uranium 235 from uranium 238, where there was a three-neutron mass difference. But with plutonium 239 and 240 there was a one-neutron mass difference, making separation very much harder. The only other suggestion was "implosion." This idea had been floating around the laboratory from the beginning but Oppenheimer had dismissed it as impractical. Here one takes advantage of the fact that the critical mass decreases as the square of the density increases. If a sample of plutonium could be made twice as dense, then the critical mass would drop by a quarter. You could in principle start with a subcritical sphere of plutonium and make the same amount of mass supercritical by squeezing the sphere. Moreover, this could be done in a very short time. Suppose the sphere has a radius of 5 centimeters and you want to shrink its volume by a factor of two, thus doubling its density and reducing the critical mass by a factor of four. This means shrinking the radius by a factor of approximately 1.3, which amounts to shrinking it by less than 2 centimeters. This can be done, if it can be done at all, in microseconds rather than the milliseconds required in the gun assembly. Thus the spontaneous fission problem would be solved. But can you do it?

The news about plutonium 240 was reported to Oppenheimer in the spring of 1944, and when he realized that the only way out was the largely unexplored possibility of implosion, he thought of resigning as director. I am persuaded that if he had done this, there would have been no atomic bomb in the war. I am also persuaded that there would eventually have been an atomic bomb somewhere. But history is written only once. On July 4 there was a meeting of all the technical personnel at the laboratory, and it was decided to devote most of the lab's activities to implosion. There was still a small group that finished the design of the uranium gun assembly device. But that was considered elementary once a sufficient amount of highly enriched uranium had been manufactured. This is something to keep in mind when one considers proliferation. If nonstate actors can get enough highly enriched uranium, they can very likely make some sort of explosive device.

At first the Los Alamos people tried to implode things like hollow cylinders. The results were awful—just mangled metal. They were not, of course, using plutonium, of which there was precious little. One of Oppenheimer's students, Robert Christy from Berkeley, investigated theoretically imploding a solid sphere and was able to show that it might be better than a spherical shell. The laboratory focused on this, which got the code name the "Christy gadget" or just the "gadget." The idea of explosive lenses was introduced. It had been brought to Los Alamos by the British physicist James Tuck and then perfected by John von Neumann. A magnifying glass focuses light because the light in the glass moves more slowly than the light in air, which bends the light beam. This is how light is focused in a magnifying glass. The same idea was used for explosives. A fast charge would ignite a slower one, and this would tend to focus the blast wave. The initial charges had to be set off almost simultaneously at several points on the surface of the sphere. The gadget had thirty-two such points and they were set off nanoseconds from each other—a virtuoso bit of electronics. The bomb itself was in two plutonium hemispheres, which had to be joined precisely. Smith made the final adjustments at Alamogordo.

One sometimes reads about the "secret" of the atomic bomb. There is no one secret but there are hundreds and hundreds of engineering details. If one insisted on pointing to a few secrets I would put on the list the use of gallium-alloyed delta-phase plutonium and explosive lenses. Because of their spies the Russians knew about these things almost from the time of their discovery. David Greenglass provided details about the explosive lenses, and Theodore Hall about some aspects of plutonium. But neither of them could be compared to Klaus Fuchs. People who were at Los Alamos have said that after Oppenheimer, Fuchs knew the most about the work at the laboratory. He also had a photographic memory. In October 1945 a Colonel Vasilevskii, presumably an intelligence agent, sent Beria a remarkable letter, which the Russians released a few years ago. Here are a few bits of it:[1]

2. Active Material

The element plutonium of delta-phase with specific gravity 15.8 is the active material of the atomic bomb. It is made in the shape of a spherical shell consisting of two halves, which just like the outer spherule of the initiator, are compressed in a nickel-carbonyl atmosphere. The outer diameter of the ball is 80–90 mm. The weight of the active material including the initiator is 7.3–10.0 kg. Between the hemispheres is a gasket of corrugated gold of thickness 0.1 mm, which protects against penetration of the initiator by high-speed jets moving along the junction plane of the hemispheres of active material. These jets can prematurely activate the initiator.

It does not mention the gallium alloy, but this may have been mentioned in other communications. The mention of the "initiator" is significant. Neutrons must be released at precisely the time at which a critical mass has been assembled. In the center of the plutonium sphere is placed a potential neutron emitter, which can be activated when the shock wave of the implosion arrives. The letter explains:

The initiator works as follows. The shock, directed towards the center, from the explosion of the outer layer of explosive is transmitted through the aluminum layer and tamper, through the layer of active material onto the surface of the hollow beryllium spherule of the initiator. The resulting stresses fracture this spherule along the planes passing through the apex of the wedge-shaped grooves, thus exposing the beryllium of the hollow spherule to the action of the alpha-particles emerging from the polonium coating on the central spherule of the initiator. This produces a neutron flux. The adjacent surfaces of the grooves collide, as a result of which the "Munroe jet" is generated, which penetrates through the thin layer of polonium and gold into the central spherule, thus putting in contact the polonium on the inner surface of the hollow beryllium spherule with the beryllium of the solid one. This also produces a neutron flux.

The neutron flux produced in the initiator attacks the active material.

Here is another important fragment of the letter.

3. Tamper (Moderator)

The tamper is a spherical shell with outer diameter 230 mm, made from uranium metal. There is an opening in the ball for inserting the active material into the interior. The opening is closed with a plug, also made of uranium metal.

The purpose of the tamper (moderator) is that it reduces the amount of active material necessary for making the atomic bomb.

The outer surface of the tamper is covered with a layer of boron, which moderates the thermal neutrons emanating from the radioactive materials of the system and are capable of causing premature detonation.

The purpose of the so-called "tamper" is threefold. On the one hand, if it is made of uranium—even uranium 238—some neutrons from the interior (the "pit") will produce additional fissions. But also it can reflect escaping neutrons back into the interior, where they can cause more fission; or it can slow down the expansion of the explosion. When an atom bomb explodes, the initial temperatures produced are greater than the interior temperature of the sun. Everything is vaporized—including the plutonium—and the gas bubble begins to expand very rapidly. As it expands, its density decreases, and after a few hundredths of a microsecond the fission reactions are turned off. The plutonium nuclei are too widely spaced. The longer this can be delayed, the more efficient is the bomb. The Hiroshima bomb was about 1 percent efficient, which meant that of the 50-odd kilograms of uranium 235, only about 1 kilogram was fissioned. The rest went off into space. The critical mass of an untampered plutonium sphere is about 16 kilograms. The tampered sphere of the Nagasaki bomb used about 6 kilograms of plutonium and was about 20 percent efficient.

I bring all this up because if the Israeli intelligence officer I quoted at the beginning is looking for "knowledge" about nuclear weapons, he need look only as far as the Internet. In Part III, I will discuss what we know about Iran's nuclear program.

PART III

Dual Use

7

Unintended Consequences Redux

FIGURE 7.1 SHOWS the fragment from the fifth draft of President Eisenhower's December 8, 1953, "Atoms for Peace" speech delivered to the United Nations General Assembly. The speech was mainly written by Eisenhower's aide General C. D. Jackson. The handwriting is Eisenhower's. Members of his cabinet, such as his secretary of state, John Foster Dulles, strongly opposed the speech, which he gave anyway. This proposal is absolutely remarkable both for what it says and what it does not say.[1]

It is unclear to me if Eisenhower had any technical background in nuclear energy. He did have some very good advisors, such as I. I. Rabi, who had gotten to know him when Eisenhower was briefly the president of Columbia University. There was also the Harvard chemist George Kistiakowsy, who had headed the explosives division at Los Alamos during the war. I do not know what exactly Eisenhower

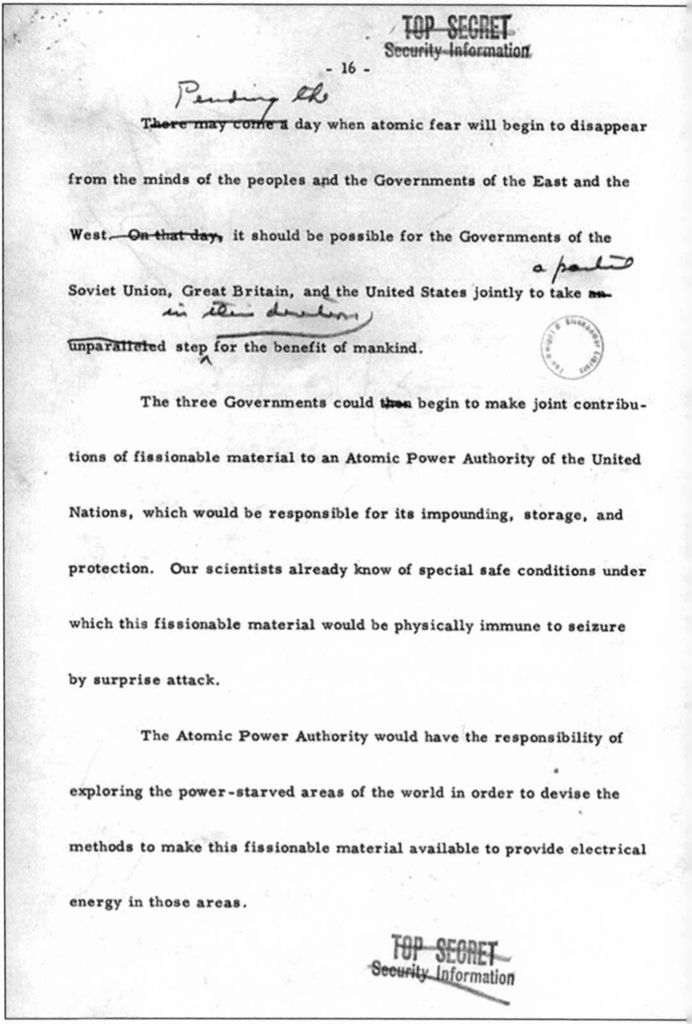

TOP SECRET
Security Information

- 16 -

~~There may come a~~ [Pending the] day when atomic fear will begin to disappear from the minds of the peoples and the Governments of the East and the West. ~~On that day,~~ it should be possible for the Governments of the Soviet Union, Great Britain, and the United States jointly to take ~~an~~ [a partial] ~~unparalleled~~ step [in this direction] for the benefit of mankind.

The three Governments could ~~then~~ begin to make joint contributions of fissionable material to an Atomic Power Authority of the United Nations, which would be responsible for its impounding, storage, and protection. Our scientists already know of special safe conditions under which this fissionable material would be physically immune to seizure by surprise attack.

The Atomic Power Authority would have the responsibility of exploring the power-starved areas of the world in order to devise the methods to make this fissionable material available to provide electrical energy in those areas.

TOP SECRET
Security Information

FIGURE 7.1. A page from the fifth draft of Eisenhower's 1953 "Atoms for Peace" speech.
(Eisenhower Presidential Library and Museum)

meant by the term "fissionable material," but one can make an educated guess. He must have meant uranium, because plutonium did not have the dual use of producing electricity. In any event it was relatively scarce and what there was, was needed for nuclear weapons. At this time reactors, especially research reactors, were largely powered by low enriched uranium. But they were starting to be converted to highly enriched uranium fuel. There was an excess of this from the weapons program, because it was using mainly plutonium. Whether Eisenhower meant LEU or HEU for his "bank" is not stated. In any event, the Russians turned the idea down, thinking it was some kind of trick to get them to give up some of their stockpile. But this did not discourage Eisenhower. In 1955 he attended a very large Atoms for Peace conference in Geneva. A swimming-pool-type reactor using LEU had been constructed for the occasion by the Oak Ridge National Laboratory. In this sort of reactor, which is widely used for research, the fuel elements are submerged in a pool of usually ordinary water. The water serves three purposes—it acts as a moderator, a coolant, and a shield against radiation. One can stand overlooking the pool as Eisenhower is shown doing in Figure 7.2 and not get irradiated.

When such a reactor is operating, there is an uncanny blue light emitted. This is what is called "Cherenkov radiation," named after the Russian physicist Pavel Cherenkov, who first observed it in 1934. It is related to the fact that the speed of light in water is roughly three-quarters of the speed of light in air. Very fast-moving electrons emitted, for example, in the beta decay of the fission products, can move faster than the speed of light in water. When this happens they emit radiation that is more intense in the blue than, say, in the red or green. This is what accounts for this effect in swimming pool reactors, and when you first see it, it is hard to believe.

One of the ideas of the Atoms for Peace program was to give grants to developing countries to help them buy research reactors. By this time the reactors were fueled with highly enriched weapons-grade uranium. As far as I can tell, no one was concerned about proliferation. Iran and Pakistan both got reactors—I would imagine that the idea of either eventually building a bomb probably seemed absurd. Not many years earlier Oppenheimer had had a conversation with President Truman. Truman asked Oppenheimer when he thought the Russians might get the bomb. When Oppenheimer said that he did not know, Truman said that he did—"never." The

FIGURE 7.2. Eisenhower at
Oak Ridge National Laboratory.
(Department of Energy, Oak Ridge National Laboratory)

Russians tested their first fission bomb—a clone of the Nagasaki bomb—in August 1949. This was followed by a British test in March 1952 and then a French test in February 1960. The Chinese first tested in October 1964. It took until May 1974 before the Indians tested and then 1998 for the Pakistanis. The latest entries in the "club" are the North Koreans. None of the people advising Eisenhower in the mid-1950s, when reactors fueled by highly enriched uranium were being handed out, seemed to think that this was a potential problem. Before I turn to what happened in Iran, I want to discuss briefly another aspect of this program—the training of nuclear scientists.

A particularly striking example is that of Munir Ahmad Khan, no relation to Abdul Qadeer. Munir was born in 1926 in Kasur, which was then British India but became incorporated into Pakistan in 1947. He studied physics and mathematics in Pakistani institutions. In 1951, on American scholarships, he completed his education at North Carolina State University and Illinois Institute of Technology. In 1957, as part of the Atoms for Peace program, he worked on reactor physics at Argonne National Laboratory. In 1958 he began working as a reactor specialist for the International Atomic Energy Agency in Vienna. Because of his work he was able to learn about the development of India's weapons-oriented nuclear program and in 1965 he was able to inform the foreign minister of Pakistan, Zulifkar Ali Bhutto, about it. After the fall of East Pakistan in 1971, Bhutto, now president of Pakistan, appointed Khan chairman of the Pakistan Atomic Energy Commission. It became his job to create the infrastructure needed for making a nuclear weapon. This included facilities for mining and converting uranium into uranium hexafluoride gas, and a facility for enriching uranium with centrifuges. It was this second facility that A. Q. Khan took over, essentially muscling Munir out. But it was Munir who was really responsible for the successful nuclear weapons program. He died in Vienna in 1999, having witnessed the first Pakistani test, for which A. Q. Khan took much of the credit. Munir was the real father.

My interest here is the beginning of the program—the first Pakistani reactor. It turns out that it is a twin of the first Iranian reactor. When the Atoms for Peace program was founded, it was decided to allow American firms to build and sell reactors abroad. One of the companies that got involved was American Machine and Foundry Atomics. American Machine and Foundry—AMF—was better known for its bowling alleys and motorcycles, but the Atomics division was created to take advantage of the commercial possibilities of nuclear energy. In March 1962 the government of Pakistan and the International Atomic Energy Agency signed an agreement in Vienna that spelled out the conditions under which AMF could supply a reactor to Pakistan. It is specified that it should be a 5-megawatt pool-type research reactor operated by the Pakistan Institute of Nuclear Science and Technology in Rawalpindi. It also allocated to Pakistan the enriched uranium required to run the reactor. Plutonium is mentioned, although I cannot imagine why that would be needed. It specifies that none of this is to be used for any military purpose and allows for Agency inspections. The reactor went critical in 1965. One interesting feature is the architecture. Munir had a classmate in Pakistan named Abdus Salam, who later went on to win a Nobel Prize in Physics. The two of them traveled to the United States to sign a contract with the architect Edward Durrell Stone. This resulted in a remarkable structure that looked more like a shrine than a site for scientific activity. I visited it in 1969 when I spent a semester as a Ford Foundation Visiting Professor in Islamabad.

Now to Iran.

In 1953, after a very troubled series of events, the CIA helped with a coup to oust the democratically elected prime minister, Mohammed Mossadegh. He had had a running battle with Shah Reza Pahlevi. Among other things, he had wanted to nationalize the oil industry. The Shah was very interested in Western technology, and in 1965 the Tehran Nuclear Research Center located in Amir Abad, a suburb of Tehran, became operational. It was the locale for the clone of the Pakistan reactor—the Tehran Research Reactor (TRR)—and it went critical in 1967. It was like its clone fueled with highly enriched uranium, over a hundred kilograms of it. In a few years the uranium 235 in the reactor fuel was largely consumed. The fission of a uranium 235 nucleus is an irreversible process that produces neutrons and fission fragments, none of which are fissionable. If there is a substantial percentage of uranium 238 in the fuel, some of this is converted into plutonium, which is fissionable and extends the lifetime of the fuel elements. But with the original configuration of the TRR with its highly enriched uranium there was not that much plutonium conversion. By 1979 the Iranians had decided not only to replace the fuel elements but to replace the TRR with a different type of research reactor altogether—the TRIGA. This type of reactor has an interesting history, which can be traced to Frederic de Hoffmann and Freeman Dyson.

De Hoffmann was born in 1924 in Vienna and grew up in Prague, where he went to school. He clashed with the young Nazis he encountered and in 1941 came to the United States, where he continued his studies of physics at Harvard. In 1944, before he had gotten his degree, he was recruited for Los Alamos. He apparently made a favorable impression on Edward Teller, and after he got his PhD from Harvard in 1948 he returned to Los Alamos to work with Teller on the hydrogen bomb. He remained there until 1955, when he became associated with the General Dynamics Corporation in San Diego. Apparently John Jay Hopkins, who was then president of General Dynamics, saw in de Hoffmann the perfect person to create a division of the company, General Atomics—GA—to be devoted to the commercial exploitation of nuclear energy. I think it is fair to say that de Hoffmann was a better entrepreneur than he was a scientist and that as the former he was very good indeed.

Probably based on his Los Alamos experience, de Hoffmann devised a modus operandi for his fledgling company. He would begin with theoretical physicists and perhaps other kinds of scientists who did not need laboratory facilities, of which GA had none. In fact the only facility it had was a small office in San Diego. De Hoffmann realized that the kind of people he was interested in had very good faculty jobs in prestigious universities. There was no way he was going to be able to persuade such people to join GA on a permanent basis. But they might be persuaded to spend a summer. They could earn a decent or more than decent consulting fee. One of the more waggish invitees kept referring to GA as "Generous Atomics." They could live in La Jolla near the Wind and Sea beach. In the summer of 1956 he assembled quite a group, a group that included Freeman Dyson and Edward Teller. De Hoffmann rented an abandoned red schoolhouse, which of course became known as the Little Red School House.

De Hoffmann proposed that the group be split into three subgroups: "test reactor," "ship reactor," and "safe reactor." Teller headed the safe reactor group and Dyson joined it. Dyson was not a nuclear physicist and knew essentially nothing about reactors. There were lectures in the morning, and the afternoons were devoted to one of the assigned problems. Teller defined the mission of his group—to design a reactor that was "intrinsically safe." High school students should be able to play with one and not get hurt. He had explained to de Hoffmann that if they succeeded, it would have commercial possibilities, because they could advertise it as being safer than anyone else's reactor. Sometime during that summer Dyson came up with the key idea.

To understand it, let us consider a swimming pool reactor in which the pool acts like both a coolant and a moderator. Suppose something happens to cause the fission reactions in the fuel elements to begin to run away. They will become heated and even threaten to melt down. The water will begin to evaporate, so some of the coolant will be lost. The natural thought is to add more water and thus cool down the fuel elements. The problem is that this added water will be relatively cool, and so it will moderate the heated neutrons from the fuel elements and cause an increase in the fission rates. This will make the situation worse. Something like this happened in the 1986 Chernobyl reactor accident. When the fission reactions began to run away, the reactor operators inserted control rods. These are made of neutron-absorbing elements like boron or cobalt. What the operators did not appreciate was that there was also graphite used in the reactor. The relatively cool graphite acted as a moderator and there was a transient increase in the fission rates—just what one did not want. Dyson asked, Suppose you could include part of the moderator in the fuel elements themselves? This would mean that if the fuel elements began to heat up, so would this part of the moderator. The heated moderator would stop moderating, and this would turn off the fission reactions, which is what one wants. This is the principle of the TRIGA, which stands for Training, Research, Isotopes, General Atomics.

As is usual in such matters, the devil is in the details. At GA there was an Iranian metallurgist named Massoud Simnad who came up with the actual amalgam that worked. It is an alloy of uranium hydride and zirconium hydride. It contains a good deal of hydrogen, which is the moderator. In June 1959 there was a licensed, prototype TRIGA that resided in one of the new buildings located on the new GA campus. I was there then as a consultant. I got $35 a day, which I thought was a princely sum. I was working in the new library one morning when I looked up to see the benevolent figure of Niels Bohr standing over me. "You have a nice library," he said as he turned to leave. De Hoffmann had gotten him to come to San Diego from Copenhagen for the TRIGA inauguration ceremony. At the ceremony Bohr pressed a switch that triggered the sudden release of the control rods. A meter showed that the initial power was 1,500 megawatts—the power of a substantial electrical power station. After a few thousandths of a second the TRIGA mechanism took over and the meter showed a half a megawatt. It worked as it had been designed to do.

By 1979 the Iranians had bought and paid for a TRIGA. But 1979 was a tumultuous year in Iran. In January the Shah was deposed and forced to leave the country. In April the Iranians voted for an Islamic republic. In November the American hostages were taken, and in December the Ayatollah Khomeni became the Supreme Leader. One of Khomeni's first acts was to declare that nuclear weapons were un-Islamic. He ordered most of the Shah's nuclear programs to be dismantled. Thousands of people were furloughed. The numbers would appear to be a statement about what the Shah's real intentions had been.

One of the survivors of the triage was the TRR, which was used for training students and making medical isotopes. In view of what was happening in

Iran, it is hardly surprising that GA lost its export license for the TRIGA. The reactor had been paid for, and one wonders what happened to the money. Clearly no other American firm was going to be able to deal with the Iranians, who turned to Argentina. But it took until 1987, after months and months of negotiation, before a $5.5 million contract was signed with the Investigaciones Aplicados. Part of this agreement was to supply a new uranium core for the TRR. The reactor was still under IAEA supervision, and this clearly would not be highly enriched uranium. In fact it was fuel elements that contained in total 115.8 kilograms of 19.75 percent enriched uranium, inserted in thirty-three metal plates. This was not delivered to Iran until 1993. The presumably degraded original 100 kilograms or so of highly enriched uranium remained in Iran. It is said to be under supervision, but by whom and where I do not know.

In Chapter 8 I am going to describe how the Shah's program was revived—something we are now living with. But I want to conclude this one with an especially odd corporate tale. In 1973, France, Belgium, Italy, Spain, and Sweden formed a company called EURODIF. This company proposed to produce low enriched uranium by the use of gaseous diffusion.

As I have mentioned previously, in this process uranium hexafluoride is forced though tiny pores in membranes and there is a small separation of isotopes as the lighter isotopes diffuse more readily. It was used at Oak Ridge to provide an essential stage that led to the highly enriched bomb uranium. It is, compared to centrifuges, a very energy-intensive method. In 1975 the Swedes pulled out, and the next year Iran acquired the Swedes' 10 percent in an agreement with France that led to the formation of a new company, SOILDIF, whose only asset seems to have been a 25 percent share in EURODIF, which gave the Iranians a 10 percent share. For the right to buy 10 percent of the production, the Shah lent EURODIF over a billion dollars to construct a factory. But after the revolution, the Iranians demanded and eventually got back a good deal of the Shah's money. But oddly they seem to still have their 10 percent share in EURODIF. One wonders what they are doing with it.

8

Among the Ayatollahs

IN MARCH 1974 the Shah declared that Iran's goal was to produce, within a few decades, 23,000 megawatts of electricity using nuclear power. This would mean constructing twenty or so power reactors. In this context a 5-megawatt reactor at the University of Tehran—the TRR—is very small beer indeed. After Ayatollah Khomeni became the Supreme Leader in December 1979 and declared that nuclear weapons were un-Islamic, he dismantled the Shah's program, which would seem to mean that he had doubts about its peaceful intentions. He had good reasons. Take the curious case of South Africa. In 1969 they began a program for making nuclear weapons which they said were going to be used for peaceful purposes such as mining. There is a good deal of uranium in Namibia, so they decided to go the route of enriching uranium. To this end they developed a process that has never been used before or since, at least on an industrial scale—the "stationary centrifuge" method.

As we have seen, a gas centrifuge separates isotopes by inserting the gas into a rapidly rotating cylinder. The gas then acquires rotational motion, and a force is exerted on the molecules that depends on their mass. This produces the separation. In the stationary centrifuge method the gas is forced to move at very high speeds through a curved pipe. The molecules are like race cars on a circular track. Again there is a force that depends on the mass and a separation of isotopes is possible. Why the South Africans chose this method, which is both inefficient and energy intensive, is not clear. It is also not clear how the Shah heard of it. The South Africans made no secret of the fact that they were trying to build nuclear devices—for "peaceful purposes," they said. Nonetheless, the Shah signed an agreement to finance the project. In return Iran was to get South African mined uranium, and, one would imagine, some of the enriched product. After the 1979 revolution Khomeni put a stop to this agreement. By this time the South Africans had produced about 55 kilograms of 80 percent enriched uranium, and by 1982 they had produced enough highly enriched uranium to make six bombs. As far as we know these were

GUESS WHO'S BUILDING NUCLEAR POWER PLANTS.

The Shah of Iran is sitting on top of one of the largest reservoirs of oil in the world.

Yet he's building two nuclear plants and planning two more to provide electricity for his country.

He knows the oil is running out — and time with it.

But he wouldn't build the plants now if he doubted their safety. He'd wait. As many Americans want to do.

The Shah knows that nuclear energy is not only economical, it has enjoyed a remarkable 30-year safety record. A record that was good enough for the citizens of Plymouth, Massachusetts, too. They've approved their second nuclear plant by a vote of almost 4 to 1. Which shows you don't have to go as far as Iran for an endorsement of nuclear power.

NUCLEAR ENERGY. TODAY'S ANSWER.

BOSTON EDISON EASTERN UTILITIES ASSOCIATES NEW ENGLAND POWER COMPANY
PUBLIC SERVICE COMPANY OF NEW HAMPSHIRE NEW ENGLAND GAS AND ELECTRIC COMPANIES

FIGURE 8.1. Advertisement from the 1970s paid for by American nuclear power companies.

(Wikipedia)

never tested, but neither was the uranium Hiroshima bomb, for that matter. Remarkably, in September 1989 F. W. de Klerk, who was then president of South Africa, decided to abandon the entire program and dismantle the bombs. The highly enriched uranium was to be turned over to the International Atomic Energy Agency inspectors. He made some pious remarks about South Africa joining the world community, but he must have been aware that Nelson Mandela would soon get out of prison. This must have factored into his decision. One residue of this activity was the creation of a cadre of experts, a few of whom became part of A. Q. Khan's proliferation network. One hopes they have been put out of business.

I want now to turn to the matter of power reactors—reactors whose stated use is the production of electric power. I must first begin by making the units more precise. There are two: megawatt thermal and megawatt electrical. The megawatt thermal power of a reactor is the actual thermal power—the heat energy per unit time—produced by the reactor. But we are interested in converting this heat into electricity by, say, making steam that runs turbines. About two-thirds of this heat energy is "wasted"—lost to friction and the like—in the process of making the electricity. Thus, the amount of electric power we get from the reactor is about a third of the thermal power. This is measured in units of megawatts electric. We must be careful when discussing the power capacity of a reactor to be clear which we mean. For example, the TRR mentioned in Chapter 7 was designed to produce 5 megawatts thermal. Unless otherwise specified, the reactor powers that I will quote henceforth will be megawatts electric. I will simply call them "megawatts."

As I have mentioned in a previous chapter, in 1975 the Shah authorized an agreement for a German company to begin constructing reactors close to the Persian Gulf city of Bushehr. There were to be two 1,198-megawatt pressurized water reactors. These reactors have a primary and a secondary source of light-water coolant. The primary source, which is kept under high pressure, comes in contact with the fuel elements. The high pressure raises the boiling point of the water, so it does not turn to steam. The secondary coolant water does turn into steam, and this is what turns the turbines that make electricity. This reactor runs on enriched uranium, which, as I have already mentioned, makes it less than ideal for producing plutonium unless there are technical modifications. For example, one can put a blanket of uranium 238 around the core, which will breed plutonium. The contract was for over $4 billion, and the work was to have been completed by 1981. But then came the revolution.

Work stopped on the reactors in January 1979 with one reactor 85 percent complete and the other 50 percent complete. In June the German company pulled out, claiming that they were owed about half a billion dollars. In September 1980 Iraq invaded Iran. In June 1981 the Israelis destroyed Iraq's Osirak reactor. One may wonder why the Israelis didn't go after the Iranian reactors, especially since Khomeni was then making anti-Israel comments regularly. In fact Israel was at the time one of Iran's principal arms suppliers—surreptitiously. They must have decided that Iran's reactors were not a threat. During the war, which lasted until 1988, Iraq bombed the Bushehr reactors, damaging them. It was not until 1995, when the Iranians signed a contract with the Russian company Altostrati, that work was again begun on one of the reactors. The other seems to have been abandoned. After various delays the Russians announced in the spring of 2009 that construction of the reactor was complete. They had supplied the many tons of low enriched uranium to run it and the final testing was begun in the fall of 2009. In

2010 fuel was loaded, by 2011 the reactor was connected to the grid, and in 2013 it was producing electric power. Unfortunately the reactor is sited fairly near earthquake fault lines. In 2013 there were two earthquakes that seem to have done some minor damage to the reactor. It is not clear that the Iranians have any specific plans to build more power reactors.

I think that there is general agreement that the Bushehr reactor does not pose a significant proliferation issue. I think that there is also general agreement that the Arak reactor does. This is a 40-megawatt thermal reactor that is being constructed about 150 miles south of Tehran. Everything about this reactor is suspicious. It was started clandestinely sometime before 2002. Its existence was revealed in mid-August of 2002 in a Washington, D.C., press conference by a representative of what is called the National Council of Resistance of Iran. The reactor was being constructed by a shadow corporation, the Messiah Energy Company of Tehran. This was a front for elements of the Iranian security forces. Once this project was revealed, the Iranians threw up the usual smoke screens. A 40-megawatt thermal reactor was hardly a plausible generator of electricity, so the Iranians said that it was going to

FIGURE 8.2. The Arak reactor site.
(Nanking, 2012)

replace the TRR in the manufacture of medically useful isotopes. But there is no evidence that anyone wants to dismantle the TRR. To the contrary, at least one study has been published in the open literature with detailed plans on how to upgrade it to 7 megawatts thermal from 5. There is an issue of refueling it, which I will discuss in Chapter 9.

The Arak reactor is suspicious for other reasons. It is a heavy-water reactor in which both the coolant and the moderator are heavy water. Such a reactor is more suitable for plutonium production than a light-water reactor because it uses as fuel natural uranium, which has the least possible percentage of uranium 235. As I have noted before, what is converted into plutonium is the uranium 238. Moreover, the Iranians have allowed only sporadic and inadequate inspections by the IAEA. The inspectors were not able to learn if the Iranians were going to build "hot cells." These are areas in which very radioactive objects, such as freshly extracted fuel rods, can be handled remotely. They would be necessary if one wanted to separate plutonium from the rest of the spent fuel element. The inspectors found that these fuel elements were of an odd design—anomalously long and thin. Perhaps they are designed so that they can be exchanged with fresh fuel elements while the reactor continues to run. In short, like so many things about the Iranian program—including the date when the Arak reactor is supposed to come on line, which has now been postponed—there is a lack of transparency, to put it mildly. The inspectors were never allowed to see any design documents.

Finally we may ask, supposing that the Arak reactor is devoted to plutonium production, how much could it produce in, say, a month? There is a useful rule of thumb for reactors operated with low enriched uranium. A 1-megawatt thermal reactor can produce about 1 gram of plutonium per day. Thus, the Arak reactor could produce about 1.2 kilograms per month. But some of this would be consumed in the fission cycle. Nonetheless, it seems safe to say that the Arak reactor could produce, per year, enough plutonium for about two nuclear weapons.

9

Breakout

IN 1993 the Iranians received 115.5 kilograms of 19.75 percent enriched uranium fuel elements for the Tehran Research Reactor—the TRR from Argentina. We do not know precisely how this reactor has been run. Its maximum power output is 5 megawatts thermal. How many days a year it has been run at this maximum, or how frequently it has been shut down, one does not know. It is also not known how the fuel has been managed. The core seems to contain eighteen fuel elements, each with 1.87 kilograms of enriched uranium, and five control elements, each containing 1.08 kilograms—the core total then being 39.06 kilograms. Thus, the Iranians were supplied enough uranium for more than one core change. Whether they changed the entire core every few years or some of the elements every year, I do not know. Given the type of reactor, it is possible to estimate how many kilograms of 19.75 percent uranium would be required for it to run at full power for a year. The answer is about 7.[1] Therefore, if it ran full time at 5 megawatts thermal, the total fuel the Iranians acquired in 1993 would have run out in about sixteen years, and the last year of its operation would have been 2009. If it has been running with less power or less often, the time would of course be longer. But in any case the end of its present operational cycle was in view. During this time it would have generated a few hundred grams of plutonium. This would be nowhere near enough for a bomb but plenty to study the chemistry of plutonium separation. Seaborg and his people only had milligrams. This means that the TRR fuel must be carefully supervised. In fact, now it is being manufactured in Esfahan from 20 percent enriched uranium produced by the indigenous centrifuges.

In the late summer of 2009 it appeared as if the IAEA might have been able to broker a deal that would be a win-win for all concerned. The idea was to send to Russia most of the 3.5 percent enriched uranium hexafluoride in Natanz. At room temperature, uranium hexafluoride is a powder and is easy to transport. The Russians would regassify it and use their centrifuges to enrich it to 19.75 percent. Then in powder form it would be sent to France.

The French use nuclear power to produce most of their electricity, so they have a large nuclear infrastructure. They would manufacture new fuel elements, which they would then send back to Iran to be used in the TRR. As it is, as of the fall of 2013 the Iranians had produced about 240 kilograms of 20 percent enriched uranium, a relatively small fraction of which is being used in the TRR. The accumulation of 20 percent enriched uranium is always a concern because of the relative ease of converting this to weapons-grade.

The Natanz centrifuge facility has been a concern from the beginning. It was built clandestinely in an unlikely location. Natanz is a city of about 40,000 known primarily for its pears and mountain scenery. The facility is a few miles from the town. Nothing might have been known about the centrifuges except for the same dissidents who revealed the Arak reactor. The centrifuges are located in a gigantic facility of some 100,000 square meters built 8 meters underground. It is protected by thick concrete walls and impervious to ordinary bombs. Inside there are two 25,000-square-meter halls, A and B, for the centrifuges. As of August 2013 the Iranians announced that in Hall A there were to be installed 25,000 centrifuges in 144 cascades. One unit was to contain the updated IR-2m centrifuges, of which six cascades had been installed, although apparently none were running. Fifty-four IR-1 cascades were running and producing low enriched uranium. All of these centrifuges were originally of the P1 type with aluminum rotors. They have peripheral speeds of about 350 meters per second. They have separative powers of 2–3 SWU per year, but from their actual performance they seem to be operating below the low end of the range. Up to the time of the 2013 report, a total of 9,704 kilograms of low enriched hex had been produced.

Why is this of concern? Because there is a potential for "breakout," in which the low enriched uranium could be highly enriched for weapons. Here I want to deal with the question of how much 19.75 percent enriched uranium could be produced if, say, a metric tonne—1,000 kilograms—of 3.5 percent enriched uranium hexafluoride were further enriched. I will make some plausible but speculative assumptions about the centrifuges.

The least speculative assumption is that in the centrifuging process no significant amount of uranium 235 is lost. We begin with a certain amount and we end with the same amount. At first sight one might think that this violates the object of the exercise, which is to enrich uranium. But in the centrifuge we are not creating or destroying uranium 235. We are simply moving it about. If we make this assumption, we can derive a simple equation that relates the amount of the product to the amount of feed. I will first take these to be uranium hexafluoride, and then later I will extract the uranium. Let P be the amount of uranium hexafluoride product in, say, kilograms, and let F be the amount of feed fed into the cascade, also in kilograms. Let N_P be the percentage of uranium 235 in the product, let N_F be likewise the percentage in the feed, and let N_W be the percentage in the waste or "tails." Then the equation is

$$P = F((N_F - N_W) / (N_P - N_W)).$$

For N_F we will put in the value .035, and for N_P the value .1975. But what about N_W? I will take as two sides of the range .0025 and .004. With $F = 1,000$ kilograms, I find, using the equation, the product to be 167 kilograms in the first case and 160 kilograms in the second. Not much difference.

However, one wants to insert uranium, not uranium hexafluoride, into the fuel elements. We may then ask, What is the percentage of uranium by mass in, say, 19.75 percent enriched uranium hexafluoride?

To determine this we must recall that the atomic mass of fluorine is 19 and that there are six fluorine atoms in each uranium hexafluoride molecule. Thus, the ratio we are looking for is given by

$$((.1975 \times 235) + (.825 \times 238)) / ((.1975 \times 235)$$
$$+ (.825 \times 238) + (6 \times 19)) \approx .68.$$

Therefore, 160 kilograms of 19.75 percent enriched uranium hexafluoride will yield about 109 kilograms of 19.75 percent enriched uranium. Recall that the core of the TRR contains about 32 kilograms of 19.75 percent enriched uranium. If the power is increased from 5 to 7 megawatts thermal, more uranium may be needed in the core, but it is clear that the low enriched uranium hexafluoride at Natanz contains much more than enough uranium to refuel the TRR.

In September 2009 the Iranians confirmed what American intelligence had already learned: they had a second and heretofore undeclared enrichment facility. This one was located near the sacred city of Qom. It is more than likely that the Iranians made this announcement when they did because they knew that the facility had been discovered. It had been under construction since at least 2006 and makes use of a tunnel complex on a military base. The details are somewhat sketchy, but this is what is presently believed. The facility is large enough to accommodate, in round numbers, 3,000 centrifuges. What makes this of special concern is that the Iranians announced their intention to install a new generation of centrifuge here. It will be recalled that the Iranians acquired prototype P1 centrifuges from A. Q. Khan's network. But they also acquired the plans for its successor—the P2.

The rotor in the P2 was to be made of maraging steel—a special very strong steel made with a minimum of carbon. As originally designed, the rotor was to be in sections joined by "bellows"—flexible steel parts that are very difficult to construct. The Iranians were not able to buy them or construct them, so they went instead to carbon fiber. This had the great advantage that the peripheral speed of such a rotor is so great that it can be built in one section with no bellows and still produce a very high degree of separation. In fact the P2 produces some 5 SWU per year. This is very much less than what the centrifuges produce in the large commercial establishments, but it is still something like a factor of two better than the P1. We can now make an estimate of what 3,000 of these machines can do.

Evidently for a start they can produce 15,000 SWU per year. We may use a SWU calculator to see what in the way of uranium 235 we can produce with this in a year.[2] If we assume that we begin with natural uranium hexafluoride with an enrichment of .7 percent and assume tails of .25 percent, then it requires 232 SWU to produce 1 kilogram of U-235, assuming we have enriched to 95 percent. Thus, in a year about 65 kilograms of uranium 235 could be produced at Qom. How many bombs this represents depends on the skill of the designer, but certainly more than one. That is why the facility at Qom is of concern. Some 20 percent enriched uranium fluoride has been produced there—as of August 2013 about 195 kilograms, all in the first-generation centrifuges.

At Natanz one could envision a different kind of breakout—something that is called "batch recycling." For purposes of discussion let us assume that the approximately 1.5 tonnes of 3.5 percent enriched uranium hexafluoride have remained in Natanz. We want to exploit them to make highly enriched uranium. The way we can do this with the least restructuring of the centrifuge cascades is to take the product we have and with some rearrangement of the tubes send it back through the cascades. It turns out that the best way to do this is in steps. This is advantageous because SWU are not additive. The number of SWU it takes to enrich from A to C is not equal to the sum it takes to go from A to B and then B to C. In the case at hand, the sum of the parts is less than the whole. It pays to go in steps. The first step in this example is to enrich uranium hexafluoride from .035 to .26. I focus on uranium hexafluoride and not uranium because in this procedure we feed the successive stages with uranium hexafluoride and extract uranium at the end. This stage requires, with the same assumptions as before, 17 SWU per kilogram of produced 26 percent enriched uranium hexafluoride. Let us assume that when this occurs all 8,000 P1 centrifuges are operating and that each puts out 2 SWU per year, so a total of 308 SWU per week. Using our formula above, and assuming that we begin with 1,500 kilograms, we can find out how many kilograms of 26 percent uranium hexafluoride are produced:

$$P = 1500(.035 - .00225) / (.26 - .0025) \approx 191 \text{ kilograms.}$$

This requires, then, 17×191 SWU $= 3,247$ SWU. This can be produced in about ten and a half weeks.

In the next step we go from 26 percent enrichment to 71 percent enrichment. This takes only 9 SWU per kilogram and produces a product given by

$P = 191(.26 - .00225) / (.71 - .00225)$ kilograms ≈ 70 kilograms.

This requires 17×70 SWU $= 1{,}190$ SWU, which takes a little less than four weeks to produce. To go from 71 percent to 96 percent enrichment takes 4 SWU and produces a product given by

$P = 70(.71 - .00225) / (.96 - .00225)$ kilograms ≈ 52 kilograms.

This requires 17×52 SWU $= 884$ SWU, which takes a little less than three weeks to produce. Thus, in about 17 weeks of actual running, 52 kilograms of highly enriched uranium hexafluoride has been produced. We must also allow some time—days—for changing the arrangements of the cascade. The amount of uranium 235 that can be extracted is thus multiplied by the percentage of uranium 235, which is

$(.95 \times 235) / (.95 \times 235 + .05 \times 238 + 19 \times 6) \approx .64.$

Thus, about 33 kilograms of uranium 235 are produced, enough for one bomb. If we had gone directly from 3.5 percent enriched uranium to 95 percent enriched uranium, it would require 81 SWU per kilogram. If we start with 1,500 kilograms, we end up with about 52 kilograms of highly enriched uranium hexafluoride at a cost of 4,121 SWU, which takes about thirteen weeks to produce.

These numbers are meant to be suggestive. They are based on limited information and some guesswork. For example, we do not know if the Iranians have been able to build 3,000 carbon-fiber centrifuges, because all of their centrifuge manufacturing facilities have been off-limits to inspections. But what the numbers do suggest is that if the Iranians ever throw off the international constraints, they could produce in not many months enough fissile material to begin to manufacture nuclear weapons. Any agreement that would enlarge this breakout time would be extremely helpful. In this respect one might note the malware attack that took place first in June and July 2009 and that was repeated the following spring. The infecting agent, identified as "Stuxnet," is so sophisticated that one assumes that it was the work of governments. It does not take much imagination to point fingers. It seems to have gotten to Natanz via some equipment that came from Siemens in Germany. What the virus did was to cause the control system of the centrifuges to go berserk—alternately speeding them up past their tolerance levels and then slowing them down. The centrifuges self-destructed—in this case, about 1,000 IR-1 centrifuges were destroyed before the Iranians got hold of the situation. This set back their program only for a time, and it has not been repeated since 2010.

In the summer of 2009 there was a very troubled vote for the presidency of Iran. Mahmoud Ahmadinejad was reelected despite accusations that the vote was badly tainted. Thousands of people took to the streets. Many were arrested and tried. There were accusations of torture. Ahmadinejad's opponent Mir-Hossein Mousavi was effectively silenced. But one thing must be clearly understood: all the political leaders in Iran are in agreement that in some form the nuclear program must go on. When he was prime minister in the 1980s, Mousavi was complicit in Iran's dealings with A. Q. Khan, as was Akbar Hashemi Rafsanjani, the *éminence grise* of the Iranian reform movement who was at that time the president of Iran.

Over the years Rafsanjani has said some very disturbing things. For example, he said this at Friday prayers in December 2001: "If a day comes when the world of Islam is duly equipped with the arms Israel has in its possession, the strategy of colonization would face a stalemate because application of an atomic bomb would not leave anything in Israel but the same thing would just produce damages in the Muslim world." The latter part of this statement persuades me that Rafsanjani has no conception of what a nuclear weapon is. By this time Israel must have several hundred, including quite possibly hydrogen bombs. A retaliatory raid by the Israelis would leave the Iranian cities in rubble. And here is something Rafsanjani said at Friday prayers on July 17, 2009:

> Our country should be united against all the dangers that threaten us. They have now upped their ransom demands and are coming forward to take away our achievements in the fields of hi-tech and particularly nuclear technology. Of course, God will not give them the opportunity to do so, but they are greedy. My brothers and sisters, first of all, you all know me, I have never wanted to abuse this platform in favour of a particular faction and my remarks have always concerned issues beyond factionalism. I am talking in the same manner today. I am not interested in any factions. In my view, we should all think and find a way that will unite us to take our country forward and save ourselves from these dangerous and bad effects, and the

emerging grudges. We should disappoint our enemies so that they would not covet our country.

This is from a "moderate."

In the Epilogue, I will summarize what we have learned and what it might mean for the future.

Epilogue

ON JULY 16, 2009, Gholam Reza Aghazadeh resigned his position as Iran's vice president for atomic energy. No explanation was offered then or since, and nothing more has been heard from him on the subject. It should be noted that his degrees were in accounting and computer science. He was not a nuclear engineer, and indeed there are odd technical lapses in his interview that I quoted from earlier. For example, he states that it is the light isotope that is moved to the edge of the rotor and that the heavy isotope remains in the center. He was succeeded by Ali Akbar Salehi, who has a PhD in nuclear engineering from MIT. Perhaps the Iranians wanted someone with greater technical competence, or perhaps it was political. It may not be coincidental that his departure coincided with the reelection of Ahmadinejad. When asked to comment, Aghazadeh is reported to have said, "It is not my field of interest. I can't care." Not exactly a ringing endorsement.

In his interview Aghazadeh makes reference to Natanz, where all this frenzied activity he describes is taking place. He describes the surveillance cameras and the visits by the IAEA inspectors. What he fails to say is that the Iranians did everything they could to keep this facility secret. It was only revealed by some dissidents in 2002. In 2013 there was a new dissident claim of a concealed site. This emphasizes the point that large centrifuge plants, unlike reactors, can be concealed. Reactors give off fission products that can penetrate any barriers used in an attempt to bury them. Antineutrinos are a fanciful example, but the noble gas xenon is not. The presence of an unexpected concentration of xenon in the atmosphere is pretty good evidence that somewhere nearby there is a reactor producing it. Centrifuge plants are, as the Iranian example shows, much simpler to conceal. No one knew about the vast Russian program, either, until they revealed it.

I began this book with a report on an interview I saw on television with an Israeli general who was said to be the head of Israeli intelligence. The general discussed the Iranian nuclear program and was asked what he thought was the line in the sand—the point at which the Israelis might have to act. He said that it was when the Israelis would learn that the Iranians had the "knowledge" to make a nuclear weapon. The burden of what has preceded this summary is to explain that the "knowledge" is not the issue. I am persuaded that the Iranians, who bought the same package as the Libyans from A. Q. Khan, got the same set of plans on how to make a weapon as the Libyans did. These were handed over to us when the Libyans handed over the rest of what they bought from Khan. The documents are classified, but it has been reported that some of them were in Chinese. This is plausible because Khan got the plans for a Chinese-designed weapon in exchange for centrifuge technology. It is more likely that he would sell these to the Iranians than that he would sell them the Pakistani variants, which may in fact have been in the possession of his rival, Munir Khan.

The Libyans were not able to do anything with the package they bought, which is one of the reasons they handed it over. They simply did not have the scientific and technological infrastructure. The Iranians did. I have tried to give some impression of how difficult it is to produce these centrifuges. Read Gholam Reza Aghazadeh's interview again. The Iranians bought a couple of used prototypes and some sort of plans. What comes to mind as an analogy is a scenario in which they were given a couple of used flat-screen televisions and some plans and instructed to build a factory that would turn them out by the thousands. What would they do? But it is worse than that. Given the television sets, all one has to do is to switch them on. But given the centrifuges, that is just the beginning. One must learn how to construct the cascades and how to make and deal with the very dangerous uranium hexafluoride and all the rest. The Iranians had help with their reactors. The Russians built the one in Bushehr, for example. But the centrifuges they did on their own. The question is, why?

Their repeated mantra, which began with the Shah, was that it was to develop nuclear energy to generate electricity. At the present time the Iranians have the third-largest oil reserves in the world—some 130 billion barrels. They have the second-largest reserve of natural gas in the world. One could understand a program that in a restrained and orderly way explored the future prospects for nuclear energy. But the Iranian program has all the urgency and secrecy of the Manhattan Project. Read again Aghazadeh's interview. What has building clandestine centrifuge facilities underground or in caves got to do with

FIGURE E.1. Geneva talks, October 15, 2013.
(U.S. Department of State)

creating electricity? And what has the way in which the technicians at Natanz conducted themselves—working day and night—got to do with it? Read again Aghazadeh's interview. To me there is only one way to read all this, and that is that the Iranians are trying to make nuclear weapons.

I take it as a given that the Iranians have the "knowledge" to do this. I also, and that has been the burden of what I have been trying to explain, take it as a given that, left to their own devices, within a fairly short time—a few months, say—they could produce enough fissile material for one or two nuclear weapons. What is wrong with this? The Israelis have hundreds, also developed clandestinely—why not let the Iranians have a few? There is a difference. The Iranians have a declared state policy. They declare it every Friday in their mosques when they chant "Death to Israel." If you are an Israeli, you must take this seriously. There is a point beyond which I think the Israelis will act. Can we keep this from happening?

I am more optimistic than I used to be. On June 15, 2013, Hassan Rouhani was elected president of Iran. By Iranian standards he seems to be a moderate. Of course, Khamenei has the final word. But Rouhani has reached out, even talking to President

Obama. Some of this was surely induced by the severe regime of sanctions that have been imposed on Iran, which includes a prohibition on the sale of oil and gas. But some may be due to a natural desire to rejoin the community of nations. In any event, on November 11, 2013, after weeks of negotiations, the Iranians signed an agreement with the International Atomic Energy Agency that greatly expands the domain of the IAEA inspectors. For the first time they will be able to visit the sites where the centrifuges are being made. Also they will be able to visit for the first time the plant where heavy water is being made. The Iranians will stop work on the Arak reactor and will not deploy the new centrifuges for six months. Their program will be in a kind of stasis during this trial period. The Israelis were not happy with the agreement. What they would like is for the Iranian nuclear program to disappear. They sometimes cite Libya as an example. But the whole Libyan program, such as it was, fit inside one CIA plane. As I have tried to make clear, here we are dealing with a vast infrastructure. I think that the best we can hope for is to make sure that it is being used only for peaceful purposes. That will be hard enough.

Postscript

THE NEGOTIATIONS with Iran about its nuclear program are in process as I write this. The outcome is uncertain. But in this postscript I would like to try to make the issues clear. To do this I will consider two limiting cases, neither one of which has any chance of being adopted but which serve as a useful foil for the discussion.

Case 1: Iran agrees to give up its entire nuclear program.
Case 2: Iran does not agree to give up any of its program but does agree to enhanced inspections.

To discuss Case 1 we need to specify what Iran's nuclear program is. Some of this has been discussed in the preceding chapters, but here I want to put it all together in summary form and add some new data. I begin with reactors. To characterize a reactor, it is useful to specify three elements:

1. The power output
2. The fuel
3. The moderator

I begin with the power output. This is usually measured in watts—a unit of energy produced per second. The practical units for reactors are millions of watts (megawatts) and billions of watts (gigawatts). If the reactor is used to produce electricity, then, as we have seen, two kinds of power produced are distinguished—thermal power and electrical power. The thermal power is the actual power that is produced in the fission process. This is generally converted into heat, which may boil water, making steam, which in turn runs the turbines that produce electricity. About two-thirds of the thermal energy is lost for various reasons in this process, hence the watts electric are about a third of the watts thermal. I will deal here only with watts thermal so we can compare various reactors.

The fuel for these reactors are ceramic uranium pellets placed in rods. The fissile isotope is uranium 235, so we must specify what percentage of uranium 235 there is in the pellets. In the Iranian reactors this ranges from about 3.5 percent to 90 percent, which is weapons-grade. Finally we must specify the moderator. The neutrons produced in fission move at about a tenth of the speed of light. As quantum mechanical objects they have a wave nature as well as a particle nature. If the neutrons are slowed, their wavelength increases, and this increases the probability for the production of fission. The slowing down is done by a "moderator." We shall specify the moderators for the Iranian reactors.

At the present time there are six functioning reactors in Iran, two others under construction, and plans for others. I will begin with the Tehran Research Reactor. Nominally it generates 5 megawatts of power. The fuel rods are kept in a swimming pool of ordinary water that acts as both a moderator and a coolant. At the present time the reactor uses about 19 percent enriched uranium. The Iranians are under way in producing the fuel elements for it. As I mentioned in the text, this reactor was originally powered by 93 percent enriched weapons-grade uranium. It appears that about 7 kilograms remain. These are highly irradiated and are apparently stored on site. For reference, about 50 kilograms of pure uranium 235 constitutes a critical mass. Uranium taken from a reactor makes a poor explosive.

The nerve center of the entire Iranian nuclear program is, as far as I am concerned, in Isfahan. It is here that yellow cake uranium is converted into uranium hexafluoride—hex—and the enriched hex is converted into useful solids that can be used in uranium fuel pellets for reactors and the like. The construction of the fuel elements is also done here. If this center were to stop functioning, the entire Iranian program would come to a halt. There are four functioning reactors on this site, all of them small. There is a subcritical reactor—meaning that the chain reaction going on in it is not self-sustaining— that uses graphite as its moderator. It was built by the Chinese, as were the other three reactors in Isfahan. This reactor seems to be used for training purposes. There is a zero-power heavy-water-moderated reactor that presumably uses natural uranium in its fuel elements. It is used for research on heavy-water applications. In principle all work on heavy water is prohibited, but the Iranians do it anyway. There is a light-water reactor that also is subcritical and used for training purposes. The most interesting of these reactors is a 30-kilowatt light-water reactor, which produces neutrons used to make medical isotopes. What is interesting about this reactor is that the fuel elements consist of weapons-grade uranium supplied by the Chinese. About 900 grams of this uranium is required to operate this reactor, which is not much, but the fact that this is weapons-grade is something to think about. The one functioning power reactor is at Bushehr. This also uses light water as a moderator and coolant. It is producing about 1 gigawatt of power. The fuel elements are 3.5 percent enriched uranium supplied by the Russians.

Two reactors are known to be under construction. In the text I discussed in some detail the 40-megawatt heavy-water reactor at Arak. All external observers agree that the only possible function for this reactor is to produce plutonium and that any agreement must change the configuration of this reactor. Finally there is a 360-megawatt power reactor that is being built by the Iranians with external aid. It is again a light-water-moderated reactor that will presumably use low enriched uranium.

There are two underground centrifuge facilities—at Natanz and Fordow. The Natanz facility is designed to hold some 25,000 centrifuges, some of which are of the latest design, replacing the original Pakistani models. At the time of the start of the negotiations, the unit had produced about 11,000 kilograms of hex enriched up to 5 percent. This sounds like a lot, until one realizes that operating a power reactor of reasonable size requires about

75,000 kilograms of uranium, 25,000 of which are replaced every two years. If the Iranian program is designed to fuel power reactors, it looks like a very small dog chasing a very large truck. This enrichment continues during the talks. On the other hand, the enrichment to 20 percent has been suspended as part of the interim agreement, and some of the existing stock has been downblended. At Fordow there are some 3,000 centrifuges, which have produced about 250 kilograms of 20 percent enriched uranium. This production is also frozen.

In addition to this there are production facilities for manufacturing new centrifuges and producing heavy water, as well as some uranium mines. The notion that all of this is somehow going to be made to vanish seems absurd. Equally absurd is the notion that none of it is going to be made to vanish. This would be a continuation of the present unsatisfactory situation. These are the two extremes the negotiators must find a path between.

Notes

Acknowledgments

Index

Notes

PROLOGUE

1. Jeremy Bernstein, *Plutonium: A History of the World's Most Dangerous Element* (Washington, DC: Joseph Henry Press, 2007).

2. Jeremy Bernstein, *Nuclear Weapons: What You Need to Know* (New York: Cambridge University Press, 2008).

1. ROUND AND ROUND

1. F. A. Lindemann and F. W. Aston, "The Possibility of Separating Isotopes," *Philosophical Magazine* 37 (1919): 523. There are some typos in this paper, which proves that even then you had to proofread carefully.

2. A reader who might like to see where this number comes from can consult an article of mine, "P. A. M. Dirac: Some Strangeness in the

Proportion," *American Journal of Physics* 77 (2009): 979.

3. Lindemann and Aston, "The Possibility of Separating Isotopes," 532.

4. Otto Frisch, *What Little I Remember* (New York: Cambridge University Press, 1979).

2. Frisch, Peierls, and Dirac

1. These notes can be found in Robert Serber, *The Los Alamos Primer* (Berkeley: University of California Press, 1992). Serber gives the basic mathematics. See also my article "Heisenberg and the Critical Mass," *American Journal of Physics* 70, no. 9 (2002): 911–916.

2. "Separation of Isotopes," in *Selected Scientific Papers of Sir Rudolf Peierls*, ed. R. F. Peierls and D. H. Dalitz (Singapore: World Scientific, 1997), 303–320.

3. I am grateful to Richard Garwin for supplying a very elegant version of such a calculator and for his comments.

3. Unintended Consequences

1. J. W. Beams, "High Speed Centrifuging," *Reviews of Modern Physics* 19 (October 1938): 245–263.

2. See the review articles by Stanley Whitley, *Reviews of Modern Physics* 56, no. 1 (1984). Whitley gives a wonderfully clear account of centrifuge physics.

3. I am grateful to Robert Norris for supplying me with a copy of the translation from Urdu of this interview.

4. GOD THE MERCIFUL, THE COMPASSIONATE

1. www.armscontrolwonk.com/file_download/32.

2. I would like to thank Richard Garwin for very helpful comments.

5. REACTORS

1. Henry D. Smyth, *Atomic Energy for Military Purposes* (Princeton, NJ: Princeton University Press, 1947). It is sometimes said that the report gave away the secret(s) of the atomic bomb. In fact it was very carefully written so that it gave away nothing.

2. Glenn Seaborg, *Adventures in the Atomic Age: From Watts to Washington* (New York: Farrar, Straus and Giroux, 2001), 100–101.

6. THE DELTA PHASE

1. nuclearweaponarchive.org/News/Voprosy2.html.

7. UNINTENDED CONSEQUENCES REDUX

1. I would like to thank Tom Cochran, Stirling Colgate, and Ken Ford for very useful comments on this chapter.

9. BREAKOUT

1. I thank Tom Cochran for providing this argument.
2. The calculator I use was kindly supplied by Richard Garwin.

Acknowledgments

I have been very fortunate to be able to consult with a group of colleagues who know much more than I do. I affectionately refer to them as my "coven." In alphabetical order they are David Albright and the ISIS website, Lowell Brown, Tom Cochran, Norman Dombey, Freeman Dyson, Ken Ford, Richard Garwin, Pervez Hoodbhoy, Robert Norris, Bill Press, Cameron Reed, Carey Sublette, Jim Walsh, and Peter Zimmerman. As T. S. Eliot wrote in his dedication to Ezra Pound in *The Waste Land*, "Il migliore fabbro."

Index